John Gilbert Baker

A flora of the English Lake District

John Gilbert Baker

A flora of the English Lake District

ISBN/EAN: 9783337271640

Printed in Europe, USA, Canada, Australia, Japan

Cover: Foto ©berggeist007 / pixelio.de

More available books at **www.hansebooks.com**

A FLORA OF THE ENGLISH LAKE DISTRICT

BY

J. G. BAKER, F.R.S., F.L.S.

LONDON
GEORGE BELL & SONS, YORK STREET
COVENT GARDEN
1885

Yet happier in my judgment, even than you,
With your bright transports, fairly may be deemed,
The wandering Herbalist—who clear alike
From vain and, that worse evil, vexing thoughts,
Casts, if he ever chance to enter here,
Upon these uncouth forms a slight regard
Of transitory interest, and peeps round
For some rare floweret of the hills, or plant
Of craggy fountain ; what he hopes for wins,
Or learns, at least, that 'tis not to be won ;
Then, keen and eager, as a fine-nosed hound
By soul-engrossing instinct driven along
Through wood or open field, the harmless man
Departs, intent upon his onward quest.
                                   WORDSWORTH.

PREFACE.

THIS work is not put forward as a completed Flora of the Lake District. I have never lived within its boundaries, but when I belonged to Yorkshire, now twenty years ago, and was engaged upon 'North Yorkshire' and the 'New Flora of Northumberland and Durham,' I several times visited the Lakes and made notes upon the plants and their range in altitude, and spent the wet days in codifying the widely-scattered records of my forerunners in the botanical exploration of the district. My collection of notes has been from time to time lent to various botanical friends who have visited it, and they have entered into the books their own observations, and the Rev. W. W. Newbould has given me his ever-ready help in copying out at the British Museum the stations from some of the books to which I could not get access elsewhere. Lately, in editing the catalogue of species for Westmoreland and Cumberland for the new edition of Watson's 'Topographical Botany,' and reading the detailed diary by Mr. Watson of his excursions in 1835, from Keswick, Kendal, and Shap, which has come into my hands as his executor, my interest in the subject has been revived, and as it does not seem likely at present to stand in the way of anything more complete, I have thought it best to publish my collection of notes as they stand, especially as the local Societies are now training up a new generation of observers, who cannot obtain access to the botanical periodicals and older works of a general character, and, even if they could, have no fair chance of understanding the gap which, from a scientific point of view, separates Watson and Borrer from Richardson and Hutton. I have considered the district which I have included as extending northward and eastward to Allonby, Wigton, Penrith, and Tebay, but have not always strictly kept to an exact limit in those directions. Broadly speaking, there are two wide tracts of country included in Watson's Lake Province not here dealt with, the low-lying northern half of Cumberland, often called the Plain of Carlisle, and the western slope of the Pennine Chain through Cumberland and Westmoreland. The Lake District, as here treated, is a mountainous tract with a distinct physical individuality of its own, and with a distinct botanical individuality, both in respect of the plants that are present and those that are rare or absent, the details of which I have endeavoured here to record as faithfully as I could. So many people have botanised at one time or another at the Lakes, that I doubt not this record will fall into the hands of many who will be able materially to modify it and add to it, and I shall be glad to receive any notes on the further range of species, with a view of using them in a new edition.

J. G. BAKER.

KEW HERBARIUM,
   Feb. 1885.

CONTENTS.

INTRODUCTORY EXPLANATIONS, *Page* 1

ORDERS.	PAGE	ORDERS.	PAGE
Ranunculaceæ,	15	Portulaceæ,	96
Berberaceæ,	23	Scleranthaceæ,	97
Nymphæaceæ,	24	Grossulariaceæ,	97
Papaveraceæ,	25	Crassulaceæ,	99
Fumariaceæ,	27	Saxifragaceæ,	101
Cruciferæ,	28	Araliaceæ,	104
Resedaceæ,	40	Cornaceæ,	105
Cistaceæ,	40	Umbelliferæ,	105
Violaceæ,	41	Loranthaceæ,	113
Droseraceæ,	43	Caprifoliaceæ,	113
Polygalaceæ,	44	Rubiaceæ,	115
Caryophyllaceæ,	44	Valerianaceæ,	118
Linaceæ,	53	Dipsaceæ,	119
Malvaceæ,	54	Compositæ,	120
Tiliaceæ,	55	Campanulaceæ,	141
Hypericaceæ,	56	Ericaceæ,	143
Aceraceæ,	58	Ilicaceæ,	146
Geraniaceæ,	59	Jasminaceæ,	147
Balsaminaceæ,	64	Apocynaceæ,	147
Oxalidaceæ,	64	Gentianaceæ,	147
Celastraceæ,	64	Polemoniaceæ,	149
Rhamnaceæ,	65	Convolvulaceæ,	149
Leguminiferæ,	66	Solanaceæ,	150
Rosaceæ,	75	Scrophulariaceæ,	151
Onagraceæ,	91	Orobanchaceæ,	157
Haloragiaceæ,	94	Verbenaceæ,	158
Lythraceæ,	95	Labiatæ,	158
Cucurbitaceæ,	96	Boraginaceæ,	166

Orders.	Page	Orders.	Page
Pinguiculaceæ,	170	Liliaceæ,	200
Primulaceæ,	171	Dioscoreaceæ,	204
Plumbaginaceæ,	174	Melanthiaceæ,	204
Plantaginaceæ,	175	Hydrocharidaceæ,	204
Chenopodiaceæ,	176	Alismaceæ,	205
Polygonaceæ,	178	Potamaceæ,	206
Thymelæaceæ,	182	Lemnaceæ,	209
Asaraceæ,	183	Araceæ,	210
Empetraceæ,	183	Typhaceæ,	210
Euphorbiaceæ,	184	Juncaceæ,	211
Urticaceæ,	185	Cyperaceæ,	214
Amentiferæ,	186	Gramineæ,	224
Coniferæ,	192	Filices,	237
Orchidaceæ,	193	Lycopodiaceæ,	244
Iridaceæ,	199	Marsileaceæ,	246
Amaryllidaceæ,	199	Equisetaceæ,	246

Postscript, 248

Index of Scientific Names, 249

Index of English Names, 255

A FLORA OF THE ENGLISH LAKE DISTRICT.

INTRODUCTORY EXPLANATIONS.

Names and Species-limits.—There are at the English Lakes just 50 ferns and nearly 850 flowering plants that are thoroughly wild, and we may count 100 more if recent introductions be included in the estimate. In the sequence, nomenclature, and limitation of species, I have followed Watson, so that in these points the present work is uniform with 'Cybele Britannica,' 'Topographical Botany,' and the earlier editions of the London Catalogue, and also with my own former books on the botany of the north of England, 'North Yorkshire,' and the 'New Flora of Northumberland and Durham.' The numbers have been widely used in the distributions of the London Botanical Society and the Botanical Exchange Club, but they are changed in the last edition of the London Catalogue. In this work I have only numbered the species that have a reasonable claim to be regarded as wild plants of the Lake district. A break in the regular sequence of the figures consequently indicates that plants that grow wild somewhere else in Britain are not found at the Lakes.

Classes of Citizenship.—These are to be understood as used in the same sense as in Watson's 'Cybele Britannica,' where they are fully defined and explained. By a '*Native*' is meant a plant which, so far as present appearances show, has established itself quite independently of man's intervention. By a '*Colonist*' is meant a well-established weed of corn-fields and arable land; by a '*Denizen*,' a plant that now looks quite wild, but may perhaps have been originally introduced by human agency; and by an '*Alien*,' a species established less thoroughly, which, without doubt, has strayed from cultivation. The fault of botanists of small experience, or sometimes of small conscientiousness, is, that in making local catalogues they swell their lists by mixing up these *Aliens*, which are often extremely fugitive in their stations, and placing them on a level with the really wild plants. As all readers of Winch and Watson and the old series of the 'Phytologist' are well aware, in the Lake district perhaps more than anywhere else in the country, confusion has been made in this way.

Maritime and Xerophilous Plants.—A considerable number of plants grow only on the seashore. These I have indicated by inserting the word 'Maritime' after the class of citizenship. A certain number of others are almost restricted to the lower limestone hills which surround on all sides the central slate mountains of the Lake district. These latter are catalogued as 'Xerophilous.'

Types of Distribution.—The types of distribution, as worked out by Watson, furnish a ready means of indicating the distribution of species through Britain as a whole. The types are as follows, viz.:—

1. *British Type.*—Species which are spread at shore-level through the length and breadth of the island.

2. *English Type.*—Species which have their headquarters in the south of England, and become rare and run out in the north of England or south of Scotland.

3. *Germanic Type.*—Species that have their headquarters in the east of England, and become rare or run out altogether in the western counties.

4. *Atlantic Type.*—Species that have their headquarters in Wales and the west of England, and run out eastward.

5. *Scottish Type.*—Species that have their headquarters in Scotland, and run out in the north of England.

6. *Highland Type.*—Species that have their headquarters in the Scotch Highlands, and grow southward only amongst the high mountains of the north of England and Wales.

7. *Intermediate Type.*—Species that have their headquarters in the north of England.

8. *Local Type.*—Species too local to be classed under any of the preceding types.

The following Table therefore will show at a glance how the plants of the Lakes are spread through the rest of Britain, and how the Lake flora stands in comparison with that of the north-eastern counties :—

	Britain.	Lakes.	North Yorkshire.	Northumberland and Durham.
British,	532	532	526	532
English,	409	208	301	251
Germanic,	127	11	38	26
Highland,	120	50	32	36
Scottish,	81	54	44	57
Atlantic,	70	12	7	5
Intermediate,	37	21	33	21
Local,	49	5	11	7
Total,	1425	893	992	935

Broadly speaking, leaving out the local species, we may reduce the other types to three, viz. (1.) General (British); (2.) Austral (English + Germanic + Atlantic); and (3.) Boreal (Scottish + Highland + Intermediate); and say for Britain as a whole, 532 species are general in their distribution, 606 southern, and 238 northern, and that out of the southern species 231, and out of the northern species 125, grow in the Lake district.

Zones of Temperature and Altitude.—By 'range' is meant

range in altitude above sea-level, and by the figures 1, 2, 3, 4 the zones of altitude in which the plant grows in the Lake district. For tracing out the vertical range of species, Mr. Watson divided the surface of Britain into two 'regions,' and six 'zones' of temperature. The two regions he called 'Agrarian' and 'Arctic.' The Agrarian includes the whole surface of the island at sea-level, and as far up the hills as arable cultivation is possible. This is up to about 600 yards above sea-level in the north of England, and 400 yards in the Scotch Highlands. All above this belongs to the Arctic region, which is so called because its characteristic plants have their headquarters within the Arctic Circle, or, at any rate, in the north. These two regions he divided each into three zones: Super-agrarian, Mid-agrarian, and Infer-agrarian; Super-arctic, Mid-arctic, and Infer-arctic. Of these six zones the coldest and the warmest, the Super-arctic and the Infer-agrarian, are not represented in the Lake district, but we have all the other four. I begin to count from below, and my zone 1 corresponds to Watson's Mid-agrarian zone; my zone 2 to his Super-agrarian; my zone 3 to his Infer-arctic; and my zone 4 to his Mid-arctic. I give a few notes on the local characteristics of these four zones.

Zone 1—Mid-agrarian zone of Watson—extends at the Lakes from coast-level to a height of 900 feet upon the hills. The average annual temperature may be estimated at from 45° to 48° Fahr. Its upper limit is marked botanically by the cessation of *Ulex* and fruticose *Rubi* in the open spaces, of *Pyrus Malus* and *Viburnum Opulus* in the woods, and of *Alnus glutinosa* and *Salix fragilis* following up the streams. All the larger lakes, Windermere, Derwentwater, Ullswater, Bassenthwaite, Crummock, Wastwater, and Haweswater, are low down in this zone, and are more or less surrounded and overtopped by thick woods, sometimes planted, but often of native growth.

Zone 2—Super-agrarian zone of Watson—includes that portion of the hill-country which lies at an elevation of between 900 and 1800 feet. The average annual temperature may be estimated at 42° to 45°. Its upper limit is marked botanically by the cessation of *Pteris, Digitalis, Erica, Parnassia,* and *Pinguicula.* Above it there are no trees, either wild or planted, except a few isolated rowans and junipers on the high crags. A great many of the mountain tarns fall within the boundaries of this zone. At the Lakes there is scarcely any arable cultivation above the Mid-agrarian zone, and there are very few houses at a higher level. The notion that the little inn at the top of Kirkstone Pass, which is the highest regularly inhabited house in the Lake district, is also the highest inhabited house in the whole of England, is a local myth which is destitute of true foundation. It stands at a little under 1500 feet above sea-level, and there are many scattered farm-houses on the Pennine chain in Yorkshire, Durham, and Northumberland at from 1800 to 2000 feet. In Allendale there is a village of considerable size called Coal Clough, which stands at from 1650 to 1700 feet in elevation. One of the principal characteristics of the Lake district, from our present point of view, is that here, broadly speaking, cultivation does not reach up to the top of the Super-agrarian, but only to the top of the Mid-agrarian zone, and that consequently at the Lakes the highest localities of a crowd of plants that follow in the footsteps of man are a whole zone, or half a zone, below their proper climatic limits, and the Super-agrarian flora at the Lakes is materially smaller than in the eastern counties.

Zone 3—Watson's Infer-arctic zone—includes a mountain belt between 1800 and 2700 feet in altitude, with an average temperature of 39° to 42°. Only the two highest tarns, Red Tarn on Helvellyn, and Sprinkling Tarn on the north of Scawfell, fall distinctly within the bounds of this zone. The rest is bare hill and slate crag, where the Alpine plants, such as *Oxyria reniformis, Silene acaulis, Sedum Rhodiola, Saxifraga oppositifolia, Saxifraga nivalis, Cerastium alpinum, Hieracium alpinum* and *chrysanthum*, and *Thalictrum alpinum*, have their headquarters, and the highest springs of the main peaks, bordered with *Montia, Chrysosplenium*, and *Stellaria uliginosa*, and beds of such mosses as *Hypnum commutatum, Bartramia fontana*, and *Bryum pseudo-triquetrum*, interspersed with *Cochlearia alpina, Epilobium alsinifolium, Saxifraga aizoides*, and *Saxifraga stellaris.*

Zone 4—Watson's Mid-arctic zone—includes the hill-tops that reach over 2700 feet, that is to say, the summits of Scawfell Pike, Scawfell, Helvellyn, Skiddaw, Bowfell, Great Gable, Pillar, Fairfield, Blencathra, Grassmoor, and High Street. Here there is nothing but bare rocky hill-top, with a very scanty vegetation of any kind. The only two plants which at the Lakes are characteristic of this zone are *Salix herbacea* and *Carex rigida*, which grow on nearly all the hills just mentioned, and are the two most decidedly arctic plants of the Lake flora. A large number of the other boreal species do not ascend into this zone at all, not because they could not bear its climate, but simply because there are amongst the loose piles of heaped stones no fit stations for them to grow in. The Lakes are the only part of England in which this Mid-arctic zone is represented.

The following Table shows the statistics of the vegetation of these climatic zones for Britain as a whole, and for the Lake district as compared with the north-eastern counties:—

	Infer-agrarian.	Mid-agrarian.	Super-agrarian.	Infer-arctic.	Mid-arctic.	Super-arctic.	Total.
Britain,	1225	1070	760	293	244	111	1425
North Yorkshire,	...	948	413	126	...	...	992
Northumberland and Durham,	...	920	418	108	...	...	935
Lake District,	...	859	301	125	28	...	893

Bibliography of Lakeland Botany.—I have given the localities for the rare plants classified under their counties, C. standing for Cumberland, W. for Westmoreland, and L. for Lancashire. As an authority for localities, B. is a contraction of my own name. For critical plants a note of admiration after the name of the collector means that I have seen and compared a specimen from that locality. The following are the principal publications that relate to the botany of the Lake district, arranged in order of date :—

1688. Lawson, Thomas. The foundation of our knowledge of the botany of the Lake district was laid in a list of 150 plants and their localities, which in this year was sent to Ray by Thomas Lawson, of Great Strickland. This was used by Ray in the second edition of his Synopsis in 1696, and was printed in its entirety in Derham's Life of Ray (in 1718, p. 213), and again in the volume of Ray's Life and Letters issued by the Ray Society in 1848 (p. 197), with the old names translated into their binomial equivalents by Professor Babington. Lawson was born in 1630, was educated at Cambridge, and when of age was ordained minister of the Church of England at Rampside, in Low Furness. In 1652 George Fox visited the district and was kindly received by Lawson, who lent him for a day his church and pulpit. Fox preached in it with such effect that Lawson and many of his congregation became Quakers. He resigned his living and settled as a schoolmaster at Great Strickland, where he was much esteemed by the Lowthers and other neighbouring gentry. He died in 1691, and his grave, with a large tombstone which was erected to his memory by one of his pupils, may still be seen in a small graveyard, without any meeting-house attached to it, at Newbyhead, which belongs to the Friends. A letter which he wrote the year before his death to Dr. Richard Richardson of Bingley, offering to meet him at Settle for a botanical excursion, is printed in the Richardson Correspondence. He wrote several books of a controversial and religious character, and a later list of plants which he found is given in Robinson's 'Natural History of Westmoreland and Cumberland,' 1709. Linnæus dedicated to his memory the genus *Lawsonia*, of which the well-known Oriental Henna is the type; and Villars named after him *Hieracium Lawsoni*.

1695. Gibson's edition of Camden's 'Britannia' contains plant-catalogues drawn up by Ray, for Westmoreland at pp. 817 and 846, for Cumberland at p. 846. They are founded mainly on Lawson's notes. In Gough's edition, 1789, the Westmoreland catalogue will be found in vol. iii. p. 164, and that for Cumberland in vol. iii. p. 206.

1744. Wilson, John. 'A Synopsis of British Plants in Mr. Ray's Method,' 8vo, Newcastle-on-Tyne, contains a trustworthy record of a large number of new localities. Wilson was a man in humble circumstances, a stocking-maker, or, another account says, a shoemaker, at Kendal. He became enthusiastically interested in plants, and educated himself to such purpose that he wrote a capital book. There is a good story about him in Pulteney's 'Sketches,' of how he was once sorely tempted to sell his only cow to buy a copy of Morison's 'Historia Plantarum,' and how a benevolent lady intervened and made him a present of it. In later life he became a teacher of botany, and removed to Newcastle, where his book was published. To Lawson and Wilson we look as the fathers of Lakeland botany. Robert Brown named in Wilson's memory the genus *Wilsonia* in *Convolvulaceæ*.

1762. Hudson, William, born at Kendal 1730, died in London 1793, was the author of the first original Flora of England in which the binomial nomenclature as invented by Linnæus was applied, and consequently he was the first to give to a great many English plants the scientific names by which they are now known. The first edition of his 'Flora Anglica' was published in 1762, the second in 1778. He practised for many years as an apothecary in London in Panton Street. In the winter of 1783 his house and all his collections and library were destroyed by fire, which was believed to have been caused designedly by a servant to conceal a robbery. After this he gave up his practice and removed to Jermyn Street, where he died in 1793. His records of localities for the Lake district are mostly expressed in general terms. Linnæus named in his memory the genus *Hudsonia* in *Cistaceæ*.

1763. Martyn's 'Plantæ Cantabrigenses' contains in the appendix, pp. 102-105, a list of the plants of Westmoreland, arranged under their localities. It is entirely compiled from Lawson and Wilson.

1777. Nicholson and Burns' 'History and Antiquities of the Counties of Westmoreland and Cumberland' contains a catalogue of plants for each county.

1782. In this year William Curtis, the originator of the 'Botanical Magazine' and author of 'Flora Londinensis,' made a botanical excursion through the northern counties, an account of which will be found reprinted in the new series of the 'Phytologist,' vol. i. pp. 36, 84, and 108.

1787-1793. Withering's 'Botanical Arrangement,' second edition, 3 vols., contains a number of stations in the Lake district, contributed by Mr. T. J. Woodward of Bungay, who visited the Lakes in company with Mr. Crowe of Norwich in 1781, and a few more were inserted in this and later editions from Mr. Hall, Mr. Atkinson, and the Rev. Mr. Jackson. Woodward was one of the most eminent English botanists of the Smithian era, and after him the fern-genus *Woodwardia* was named. His contemporary, Goodenough, the monographer of the English *Carices*, was for a few years at the end of his life Bishop of Carlisle. His extensive herbarium, which was presented to Kew about 1880 by the corporation of Carlisle, did not, however, contain any Lake plants specially localised.

1794. Hutchinson's 'History of the County of Cumberland and some Places adjacent,' 2 vols., Carlisle, republished in

1805, contains a list of plants drawn up by the Rev. W. Richardson of Dacre. He seems to have derived a good deal of his information from a Keswick guide of the name of Hutton. Neither Richardson nor Hutton had the knowledge needful for the task they undertook, and this list, although perhaps it may give a useful hint sometimes to a resident explorer, is so inaccurate and untrustworthy that I have scarcely ever cited it.

1805. Dawson Turner and Dillwyn. 'The Botanist's Guide through England and Wales' contains a long catalogue of both flowering plants and cryptogamia for both counties, —for Cumberland at p. 143, and for Westmoreland at p. 638. The principal contributors of new localities are Turner himself, the Rev. John Dodd of Wigton, the Rev. W. Wood of Whitehaven, the Rev. J. Harriman of Eglestone, and Messrs. Jos. Woods of London and T. Gough of Kendal.

1818. Otley, Jonathan. In this year was published the first edition of Otley's well-known guide-book, which contains the first good map of the Lake hills, and on which the later guide-books are all more or less founded. Otley worked out a few fresh plant stations, and though, like Wilson, a self-taught man in humble circumstances, he did a great deal for the investigation of the geology and physical geography of the district.

1824. Winch, N. J., contributed to the 'Newcastle Magazine' in 1824 (vol. iii. pp. 494, 530, and 575) a list of the plants of Cumberland and their localities, which was reprinted as a separate work in 4to in 1833. Winch was a capital botanist, and the author of an excellent 'Botanist's Guide through the Counties of Northumberland and Durham,' published at Newcastle, in two volumes 8vo, in 1805-1807. The younger De Candolle named after him the genus *Winchia* in *Apocynaceæ*.

1832. 'Annals of Kendal,' by C. Nicholson, contains,

pp. 221-225, an excellent list of the plants of the neighbourhood, drawn up by Mr. T. Gough, which he kindly revised for me a few years ago.

1835. Watson, Hewett Cottrell: 'New Botanist's Guide to the Localities of the Rarer Plants of Britain,' vol. i. England and Wales; vol. ii. Scotland, and Supplement, 1837. This is substantially a new edition, brought up to date, of the 'Botanist's Guide' of Turner and Dillwyn. Mr. Watson, who died in 1881, devoted himself during a long life to the special study of the geography of British plants. His books, of which the principal are 'Cybele Britannica' and 'Topographical Botany,' extend over a period of forty years, and in the present work I have followed up his methods and used his zones of altitude and other generalisations. He stayed for some time in the Lake district, in 1833, at Keswick, Kendal, and Shap, and made copious notes on the plants, which are now preserved at Kew along with his herbarium. His observations on the altitudinal range of Lake plants are printed in the 'Cybele,' vol. iv. p. 334.

1835. Woods, Joseph, in Hooker's 'Companion to the Botanical Magazine,' vol. i. p. 298, has a long paper called 'Notes of a Tour in the North of England in 1835,' which relates chiefly to the Lakes. He had previously visited the district in 1800, 1808, and 1814. He was the author of the well-known 'Tourist's Flora,' and was the first to study carefully our indigenous roses, of which he published a monograph in vol. xii. of the Transactions of the Linnæan Society. Robert Brown named after him the fern-genus *Woodsia*.

1836-1872. Hindson, Isaac, of Kirkby Lonsdale, during these years worked at the flora of that neighbourhood. His manuscript list of localities is now in my possession.

1842-1854. In the old series of the 'Phytologist' will be found papers that relate to Lake botany, as follows:—In volume ii., at p. 316, by J. Sidebotham; at p. 424, by Borrer; at p. 375, by G. S. Gibson; and at pp. 422 and

1045, by James Backhouse, the monographer of the British *Hieracia*. In volume v., at p. 26, is a paper by Professor J. H. Balfour, and at p. 1 a catalogue of Gosforth plants by Joseph Robson. In the second series there are papers of the plants of Humphry Head by Dr. Windsor, in vol. ii. p. 257, and Mr. C. J. Ashfield, in vol. v. p. 257.

1843. Jopling's 'Sketch of Furness and Cartmel' contains a list of the plants of the district, drawn up by Messrs. Aiton and Wilson.

1855. Harriet Martineau's 'Guide to the Lakes' contains two lists of plants, one for Windermere and its neighbourhood, drawn up by Mr. F. Clowes, and one for Cumberland, drawn up by Mr. W. Dickinson. Mr. Clowes, who has long been in practice as a surgeon at Bowness, knows the Windermere district most thoroughly, and has kindly annotated my notebook. Mr. W. Dickinson was a land-surveyor, who had lived in the west of Cumberland all his life, and knew every part of it most thoroughly. He was a man, Mr. Hodgson tells me, of robust constitution, striking appearance, and a first-class pedestrian, ardently fond of natural history in every department, especially botany and ornithology. He made a large number of careful drawings of Lake plants. Besides this list he also published a Cumberland glossary, and a book called 'Cumbriana,' a collection of humorous anecdotes of the yeomen and labourers of the county. He died in 1882 at his home at Thorncroft, near Workington, at the age of eighty-three. The beautifully illustrated work on 'The Lake Country,' by Mrs. E. Lynn Linton, and Black's Guide, also make mention of a few additional localities.

1865. Linton, W. J. 'Ferns of the English Lake Country,' with numerous original woodcuts, contains in a cheap, handy form a description of all the Lake ferns, and a list, drawn up by Mr. Clowes, of the numerous varieties named by Moore, Lowe, and others, which have been found in the district. A new edition, edited by Mr. J. M. Barnes, has since been published. Aspland's 'Guide to Grange' contains a list of the ferns and rare plants of that neighbourhood, drawn up by Mr. A. Mason.

1874. Hodgson, Miss Elizabeth, of Ulverstone, wrote an excellent paper, called 'North or Lake Lancashire, a Sketch of its Botany, Geology, and Physical Geography.' It appeared first in Trimen's 'Journal of Botany,' vol. iii. pp. 268 and 296, and afterwards separate copies were issued. During many years Miss Hodgson was my principal botanical correspondent resident in the district. She died at a comparatively early age, and the preparation of this paper was her recreation during a tedious illness. Before publishing it she sent up her herbarium to me to have the names of the plants verified, and afterwards presented her collection to the British Museum.

1880-1883. The botanical papers which have appeared in the 'Transactions of the Cumberland Association for the Advancement of Literature and Science' are as follows, viz.: In vol. v.—on the 'Grasses of Mid-Cumberland,' by W. Hodgson; 'Observations on the Flowering Plants of West Cumberland,' by J. Adair; 'On the Lichens of Cumberland,' by Rev. W. Johnson; and 'Contributions towards a List of West Cumberland Flowering Plants and Ferns,' by the members of the Botanical Society of Whitehaven, edited by Mr. Adair. In volume vii.—'Contributions towards a List of the Fungi growing round Carlisle,' by Dr. Carlyle; 'Notes of the Flora of Ullswater District,' by W. Hodgson; and 'Additions to the List of Flowering Plants of West Cumberland,' by J. Glaister and Dr. Leitch of Silloth. To the separate copies of his paper on the Ullswater flora, Mr. Hodgson, who I am glad to see has just been elected an Associate of the Linnæan Society, has added a full catalogue of the species found in the district that drains into the lake, and I am further indebted to him for looking through my notes before publication, and making material additions to the list of localities and local names.

Rooke Collection of Botanical Drawings. Mr. Rooke was an artist who lived at Whitehaven. He was fond of pedestrian rambles, and made sketches of every object that fell in his way. His series of drawings of Lake plants, which were executed with great fidelity, and are bound up in five or six volumes, after his death were purchased by J. C. Brown, Esq. of Hazel Holme, Whitehaven, in whose possession they still remain.

Miss Wilson's Collection of Botanical Drawings. Miss Harriet Wilson has also made a large collection of water-colour drawings of Lakeland plants, and kept careful note of their stations. She kindly lent me her collection, and I have often cited her localities for rarities.

I. FLOWERING PLANTS.

Class I. DICOTYLEDONS or EXOGENS.

Division I. THALAMIFLORÆ.

ORDER RANUNCULACEÆ.

Clematis Vitalba, L. (Traveller's Joy). Alien. Hedges and thickets; an occasional straggler from cultivation.

C. Hedge near Ravenglass.—(Whitehaven Cat.)

L. High Stott park, on the west side of Windermere.—(Miss Hodgson.) On the limestone rocks near the top of Yewbarrow, over Grange, and hedges in Carter Lane, between Grange and Kents Bank.—(W. Foggitt, F. Clowes, B.)

2. *Thalictrum alpinum*, L. (Alpine Meadow-Rue). Native. Highland type. Range 2-4. Damp places; not uncommon on the higher mountains (2000 feet).

C. Scawfell Pikes, overlapping *Salix herbacea*, Sprinkling Tarn, Styhead Tarn, and Great End.—(Watson, J. Robson.) Summit of Blacksail Pass.—(W. Foggitt.) Hanging Knott, at about 2000 feet, with *Juncus triglumis*.—(D. Oliver.)

W. Helvellyn, on the Striding-edge Rocks, at 850-900 yards, and St. Sunday's Crag.—(Woods, W. Foggitt, B.)

3. *Thalictrum minus*, L. (Lesser Meadow-Rue). Native. Scottish type. Range 1-2.

Var. *maritimum*. On the coast sandhills.

C. Eskmeals, near Ravenglass.—(J. Robson.)

W. Near Arnside Tower.—(Hindson.)

L. Plentiful on Walney Island; first recorded by J. Lawson in 1688. Biggar Bank, Walney west shore.—(Miss Hodgson.)

Var. *montanum*. On cliffs of limestone and slate.

C. Black rocks of Great End (600 yards), and ravine of the Wastwater Screes.—(Wood, Watson, Oliver.) Honister Crag. —(D. Oliver.) Piers Ghyll, Scawfell.—(Rev. A. Ley.)

W. Cliffs and débris of Scout Scar, Kendal, and in the limestone crevices of Whitbarrow and Huttonroof.—(Gough, W. Foggitt, Watson, B.) Harrison Stickle, Great Langdale. —(Rev. A. Ley.) Rocks above Mardale.—(Watson.) High Street, above Blea Water.—(Rev. A. Ley.)

3* *Thalictrum flexuosum*, Bernh. *T. majus*, Smith, non Jacq. (Common Meadow-Rue). Native. Intermediate type. Range 1. Frequent in damp places about the great lakes.

C. Shores of Derwentwater, at Lodore, etc.—(Watson, B.) North shore of Ullswater, from Glenridding down to Pooley Bridge; first recorded by E. Robson. Side woods and shore of Ennerdale Lake.—(W. Dickenson, Rev. F. Addison.) A single root at Isell.—(Rev. J. Dodd.) Vale of St. John, at Wanthwaite Bridge (*Kochii*, Bab. !)—(W. Hodgson.) Borrowdale.—(Rev. William Hind.)

W. In Patterdale, below Brothers Water, and along the south shore of Ullswater, especially at Sandwyke and Howtown.—(D. Oliver, W. Hodgson, B.) Brathay (*Kochii*, Bab. !). Banks of Haweswater Beck, at Rossgill.—(B.) Drawn from Skelwith in Miss Wilson's collection. Windermere shore, near Ferry Inn.—(W. Foggitt, B.)

L. Shore of Coniston Lake at Waterhead, and elsewhere. —(Miss Beever, Rev. F. Addison, B.) Foot of Windermere. —(Miss M. A. Ashburner.) By the Leven, at Low-wood Bridge.—(Miss Hodgson.) By the old well at Cartmel.— (T. Lawson.) I cannot distinguish between *T. flexuosum* and

Kochii. See Babington in Annals of Natural History, series 2, vol. xi. p. 265.

4. *Thalictrum flavum*, L. Native. English type. Range 1. Watery places. Rare.

C. By the Greta, in Howrayfield, near Keswick.—(L.) By the Maryport and Carlisle Railway, to the west of Dalston Station.—(W. Hodgson.)

W. Not uncommon on the shores of Windermere. (Clowes. A misprint for *T. majus.*) Near Arnside, in low marshy ground below Middlebarrow Wood.—(D. Oliver !, J. C. Melvill, C. Bailey, B.)

6. *Anemone nemorosa*, L. (Wood Anemone). Native. British type. Range 1-3. Common in woods and upon the mountains, ascending to 2200-2300 feet on Grisedale Pike (Watson); to 850 yards on Helvellyn, and to the limestone pavement of Huttonroof Crags and Farleton Knot (B.).

Adonis autumnalis, L. (Corn-Adonis, or Pheasant's Eye). Alien.

C. Grows as a weed in several cottage gardens at Aspatria, where it has probably been cultivated at some period or other. (W. Hodgson.)

11. *Ranunculus aquatilis*, L. (Water Crow-Foot). Native. British type. Range 1. Ponds and ditches; frequent in the low country, ascending to 300 yards.—(Watson.)

C. The highest station in which I have seen it is in Haweswater Beck at Rossgill.

Of the sub-species, *peltatus, floribundus, heterophyllus, trichophyllus,* and *Drouetii* all occur.

11*. *Ranunculus confusus*, Godr. Native. British type. Range 1.

L. In Windermere, near the Ferry, and in other places.— (Hiern. !)

11*. *Ranunculus circinatus*, Sibth. Native? English type. Range 1. Given in Black's Guide as a plant of Ullswater. Confirmation required.

11*. *Ranunculus fluitans*, Lam. Native. English type. Range 1.

C. Given in Black's Guide as a plant of Derwentwater. Confirmation wanted. In the bed of the Eamont, a little below Pooley Bridge; also in the bed of the Eden at Lazonby.—(W. Hodgson.)

W. In the Eden at Appleby; a characteristic drawing in Miss Wilson's collection.

12. *Ranunculus Lenormandi*, F. Schultz. Native. English type. Range 1-2. Not infrequent in bogs, ascending to 400 yards.

C. Dent Hill and other places in the Whitehaven district.—(Rev. F. Addison !, Whitehaven Cat.) Vale of Lorton, at the foot of Whinlatter.—(B.) In Borrowdale, near Seathwaite.—(W. Foggitt.) Keswick.—(Watson, Rev. F. J. A. Hort.) Near the foot of Blake Fell in Lamplugh.—(W. Hodgson.)

W. Great Langdale.—(Hiern. !) Shap.—(Watson.) Swindale.—(B.)

L. Coniston near the head of the Lake, and in other places.—(Backhouse, Miss Beever.) Drawn from Lake Bank, Coniston, in Miss Wilson's collection. Plumpton peat-trenches.—(Miss Hodgson.)

13. *Ranunculus hederaceus*, L. (Ivy Crow-Foot). Native. British type. Range 1-2. Frequent in the same sort of places as the last, ascending from the shore at Walney Island (C. Bailey); to 500 yards (Watson).

14. *Ranunculus Ficaria*, L. (Lesser Celandine or Pilewort). Native. British type. Range 1. Common in woods and meadows through the lower zone.

15. *Ranunculus Flammula*, L. (Lesser Spearwort). Native. British type. Range 1-3. Common in swampy places and about the lakes and tarns; ascending to 700 yards on Helvellyn, and to the highest springs of High Street and Coniston Old Man. Dr. Boswell, in the Exchange Club Report for 1880, p. 28, identifies a plant gathered by Mr. Bolton King on the stony shore of Ullswater with the Loch Leven *R. reptans*. Long ago Woodward recorded it from Coniston Water, but specimens gathered there lately by Mr. A. G. More were the var. *pseudo-reptans;* and I have also notes of this var. from Rydal Lake (J. C. Melvill), Ennerdale Lake (Whitehaven Cat.), Bassenthwaite Lake from Peelwyke downwards (W. Hodgson), and Urswick Tarn (Miss Hodgson).

16. *Ranunculus Lingua*, L. (Greater Spearwort). Native. English type. Range 1. Deep ditches and shores of some of the tarns. Rare.

C. Naddle Beck, near Keswick.—(M.) In West Cumberland at Cleator (Rev. F. Addison), and Ennerdale, Eskdale, and Wastdale (J. Robson). Ditches in the moss at Newton Regny.—(B.) Ditches near Dubmill, Allonby; and in Shawk Beck, near Curthwaite.—(W. Hodgson.)

W. Foulshaw Moss and other places near Kendal.—(Wilson, Gough.)

L. Mosses and damp meadows in Furness and Cartmel.—(Aiton.) In the water and ditches of the moss by Hawkshead.—(Lawson.) Esthwaite Water.—(Rev. W. Wood.) Borders of Urswick Tarn, near Ulverstone.—(Miss Hodgson.)

18. *Ranunculus auricomus*, L. (Wood Crow-Foot, Goldilocks). Native. British. Range 1. Woods and hedgebanks, frequent, ascending to 200 yards.—(Watson.) First recorded as a Westmoreland plant by Lawson in his Catalogue of 1688. A stunted variety occurs abundantly on the Cumberland shore of Ullswater, at Oldchurch.—(W. Hodgson.)

19. *Ranunculus acris*, L. (Bitter Buttercup). Native. British type. Range 1-3. Everywhere common in grassy places, ascending to 700 yards on Great Gable (B.); and to 900 yards on Helvellyn (Watson). 'Variat in Westmorlandia caule uni- vel subbi-floro et calyce valde hirsuto.'—(Hudson, Fl. Angl. ii. 241.) There is a very dwarf 1-2-flowered form with little-cut leaves on the limestone of Whitbarrow. Dr. Boswell refers a plant gathered near Westward by Rev. R. Wood to var. *vulgatus*.

20. *Ranunculus repens*, L. (Creeping Crow-Foot; local name, 'Meg wi' many feet'). Native. British type. Range 1-2. Common in ditches and grassy places, ascending to 710 yards.—(Watson.) The highest stations for which I myself have a note are Hayes Water and streams round Low Water, Coniston Old Man.

21. *Ranunculus bulbosus*, L. (Bulbous Crow-Foot, or Buttercup). Native. British type. Range 1. Common in meadows in the lower zone.

22. *Ranunculus hirsutus*, Curt. (Pale Hairy Buttercup). Native. English type. Range 1. Waste ground near the sea.

C. Workington Marsh and Drigg near Ravenglass.—(M.) Near the Old Kiln Farm, Allonby, and occasionally along the shore of Solway Firth.—(W. Hodgson.)

L. Abundant at Barrow-in-Furness.—(W. Foggitt.) Island of Walney, on the west side of Biggar Bank.—(Miss Hodgson.) In a lane near the Ferry on Walney Island and in a grassy marsh a little to the south of it.—(Dr. F. A. Lees.) Meadows and wet places near the Duddon.—(Aiton.)

23. *Ranunculus sceleratus*, L. (Celery-leaved Crow-Foot). Native. British type. Range 1. Ditches and ponds. Rare.

C. Allonby and Seascale.—(Whitehaven Cat.) Workington.—(D. Oliver.) Holme Dub, near Mealrigg, and about Dubmill, sparingly.—(W. Hodgson.)

W. Formerly abundant on Brigstear Moss, near Kendal.—(Gough.) Pools along the shore between Arnside and Milnthorpe.—(B.) Marsh below Middlebarrow Wood, Arnside.—(B.)

L. Bogs and damp places in Furness and Cartmel.—(Aiton.) Edge of a pit at Flookborough.—(C. Bailey.) Peat ditches, Ulverstone Moss.—(Miss Hodgson.) Saltmarsh west of Humphrey Head.—(B.)

25. *Ranunculus arvensis*, L. (Corn Crow-Foot). Colonist. English type. Range 1. Cultivated fields and gardens. Rare.

26. *Caltha palustris*, L. (Marsh Marigold). Native. British type. Range 1-3. Common about the lakes and tarns. The highest place in which I have seen it is at the Red Tarn, 800 yards. Mr. Watson notes it at 900 yards. The form of the plant in upland places is var. *minor*.

27. *Trollius europæus*, L. (Globe-Flower). Native. Scottish type. Range 1. Frequent about the large lakes; Derwentwater, Borrowdale, Watendlath Valley, Ullswater, Windermere, the Brathay Valley, Rydal Falls and Coniston Water. Ascends to 300 yards.—(Watson.)

C. Rare in the Whitehaven district. Meadows by the Mite at Ravenglass.—(J. Robson.) Aspatria Mill.—(Rev. J. Dodd.) Yeathouse, Eskatt, and Rowrah.—(Whitehaven Cat.) Newton Regny Moss near Penrith.—(B.)

W. Banks of the Mint near Kendal.—(Gough.) Mardale, Rossgill, and Crosby Ravensworth.—(Watson.)

L. Roadside near Dale Park in Furness Fells.—(Atkinson.) Buckbarrow and Gateside, Cartmel.—(Aiton.) Colton Beck Wood, Furness Fells, and Seathwaite Tarn Beck at Newfield.—(Miss Hodgson.)

29. *Helleborus viridis*, L. (Green Hellebore). Native. Xerophilous. English type. Range 1. Woods and hedgerows; probably truly wild on the limestone only.

C. Threepland Ghyll.—(Rev. J. Dodd.) Duddon Woods and Plumbland near Workington.—(Tweddle.)

W. About Clathrop Hall.—(E. Robson.) Two places near the terminus of the Windermere Railway.—(F. Clowes.) Wood near Arnside Knot.—(D. Oliver.) Pastures and hedges near Arnside Tower.—(Ashfield.) In a lane at Arnside, apparently in the very locality mentioned by Gerarde two hundred years ago.—(C. Bailey.) In Westmoreland it is called 'Felon Grass.'—See Wilson, Syn. p. 130.

L. Slack Woods, Grange.—(Miss A. Butler.)

Helleborus fœtidus, L. (Stinking Hellebore). Alien.

W. Near the road between Bowness and Kendal.—(F. Clowes.)

31. *Aquilegia vulgaris*, L. (Common Columbine). Native. Xerophilous. English type. Range 1-2. Woods and fields. Frequent in the central Lake country. Often truly wild, but sometimes only a stray from cultivation.

C. Islands and shores of Derwentwater, and in the Borrowdale meadows as far up as Rosthwaite.—(Winch, W. Foggitt, C. Bailey, B.) Cumberland shore of Ullswater, half a mile above Gowbarrow Hall, and near the mouth of Airey Beck.—(Balfour, W. Hodgson.) Dowthwaite, on rocks in a ravine above 1500 feet.—(W. Hodgson.) Bassenthwaite Lake.—(M.) Head of Wastwater.—(Winch, Miss Beever.) Woods at Gosforth.—(J. Robson.) Yoltenfews and shores of Crummock Lake.—(Whitehaven Cat.!) Side of the Lorton road, two miles out of Cockermouth.—(B.) Fine in fissures of rock, 1500 feet above the Vale of St. John.—(J. Backhouse.)

W. Near the Ferry, and many other places on the shores of Windermere.—(M., B.) Huttonroof.—(Hindson.) Brigstear and other places on the limestone near Kendal.—(Hudson, Gough.) Middlebarrow Wood, and other places about Arnside.—(J. C. Melvill, B.) Western slope of Loughrigg above Scroggs.—(W. H. Hills.)

L. Coniston, near the lake, and in Ghylls; not plentiful.—(Miss Beever.) Dalton-in-Furness, Rowdsey Wood, and Plumpton rocks, Ulverstone shore.—(Miss Hodgson.) Hudson mistakes the wild lake plant for *A. alpina.*—See Sir J. E. Smith, Eng. Bot. t. 29.

Delphinium Ajacis, Reich. Alien.

C. A weed in a corn-field at Dean, 1874.—(W. B. Waterfall.)

Aconitum Napellus, L. (Monk's Hood). Alien. Found by Mr. R. Lowther, near Hugh's Crag Bridge. An escape from cultivation.—(W. Hodgson.)

34. *Actæa spicata*, L. (Herb Christopher, or Bane Berries). Native. Intermediate type. Range 1.

W. Mountainous pastures above Troutbeck.—(Woodward, in Bot. Guide, ii. 644.) Rocky wood on the limestone (Whitbarrow?) between Kendal and Arnside Knot.—(Dr. F. A. Lees.) Sandwyke, Ullswater.—(Rev. W. Richardson.) 'I have never found this plant at this station.'—(W. Hodgson.)

ORDER BERBERACEÆ.

35. *Berberis vulgaris*, L. (Barberry). Denizen. English type. Range 1. Woods and hedges. Rare, and perhaps always introduced.

C. Hassness Woods, Buttermere, doubtless introduced.—(B.) Irton, Muncaster, and Ravenglass.—(Whitehaven Cat.) Hedgerow near Penruddock, as a substitute for *Cratægus Oxyacantha*, not truly wild.—(W. Hodgson.)

W. Windermere shore near the Ferry Inn.—(W. Foggitt.) Banks of Rydal Lake.—(Balfour.) Kirkby Lonsdale, not infrequent, but not truly wild.—(Hindson.) Denny Hill, below Haweswater, and hedges at Hackthorpe and Great Strickland. —(B.)

L. Between Penny Bridge and Colton Beck Bridge; doubtfully wild.—(Miss Hodgson.) Hedges by the roadside near Storrs Hall.—(B.)

Epimedium alpinum, L. (Barrenwort). Alien. Woods and hedge-banks; an occasional straggler from cultivation. Reported on old authority from Threlkeld, Cockermouth, Borrowdale, and Helvellyn. (See Phytologist, vol. ii. p. 3, and Winch, Contrib. p. 8.)

C. Wood (Irton?) by the Wastwater stream, half a mile from Santon Bridge, with *Erica vagans*.—(Borrer.)

W. Under hedges in several places about Fox How and Ambleside, no doubt introduced.—(Sidebotham.)

ORDER NYMPHÆACEÆ.

36. *Nymphæa alba*, L. (White Water-Lily). Native. British type. Range 1-2. Plentiful in most of the lakes and tarns. Derwentwater, Windermere, very fine at Grasmere, Coniston Water, Esthwaite Water, Blea Tarn, Urswick Tarn, Brotherswater, Watendlath Upper Tarn, etc. The highest station I have for it is Angle Tarn, Place Fell, 500 yards, where it was noted by Mr. W. Foggitt. Furness, in the tarn near Bigland Hall.—(Aiton.) I was once at Grasmere at the annual 'Rush-bearing,' when the whole church was hung round with white lilies from the lake, but they fade very soon.

37. *Nuphar lutea*, Sm. (Common Yellow Water-Lily). Native. British type. Range 1. Lakes and tarns; not so

common as the white water-lily. Grasmere, Thirlmere, Derwentwater, Coniston Water, etc.

C. In St. John's Beck, near Smeathwaite Bridge.—(W. Foggitt.) Loweswater, Mockerkin Tarn, and Braystones Tarn.—(Whitehaven Cat.) Dubmill, near Allonby.—(W. Hodgson.)

W. Cunswick Tarn, near Kendal.—(Gough.)

L. Ayside Tarn near Cartmel.—(Aiton.) Latterrigg Tarn, Woodland.—(Miss Hodgson.)

ORDER PAPAVERACEÆ.

40. *Papaver Argemone*, L. (Long Rough-headed Poppy). Colonist. British type. Range 1. Cultivated fields. Rare.

C. Middletown, St. Bees.—(Whitehaven Cat.) Ullswater, scarce; fairly plentiful about Bullgill Railway Station, Maryport.—(W. Hodgson.)

L. Furness shores at Roosebeck.—(Aiton, Miss Hodgson.)

41. *Papaver dubium*, L. (Scarlet Poppy). Colonist. British type. Range 1. Cultivated fields; seen only on the outskirts of the Lake country. Marked in the Catalogues for Whitehaven, Kendal, Kirkby Lonsdale, Penrith (where it ascends to 300 yards), Clifton, Clibburn, Arnside, Burton in Lonsdale, and Grange-over-Sands; not for Keswick, Ambleside, or Shap. Common in West Cumberland about Aspatria and Blennerhasset.—(W. Hodgson.)

42. *Papaver Rhœas*, L. (Common Red or Crimson Poppy). Colonist. British type. Range 1. Less common than the last in the Lake country, and not ascending so high. Marked in the lists for Whitehaven, Ullswater, Kirkby Lonsdale, Barrow and Grange-over-Sands; not for Keswick, Ambleside, or Shap. Mr. Clowes says he never saw it near Windermere.

Sandy corn-fields towards the Solway. Very rare, and likely to become extinct.—(W. Hodgson.)

Papaver somniferum, L. (White or Opium Poppy). Alien. An occasional straggler from gardens.

C. Sellafield and St. Bees.—(Whitehaven Cat.) North side of Ullswater, near a cottage.—(Balfour.) Extinct.—(W. Hodgson.)

L. Rubbish in Carter Lane, between Grange and Kents Bank.—(B.)

44. *Meconopsis cambrica*, Vig. (Yellow Welsh Poppy). Alien. A frequent straggler from gardens all through the central Lake district, but never seen, so far as I am aware, far away from houses and gardens. Mr. and Mrs. Hills incline to consider it truly wild on the shores of Windermere and Esthwaite Water. First recorded by Hudson.

45. *Chelidonium majus*, L. (Greater Celandine). Denizen. English type. Range 1. Frequent in hedges and by roadsides near villages and farm-houses. Aspatria, Keswick, Lorton Vale, Shap, Crosby Ravensworth, Colwith, Clawthorp, Townend in the Winster Valley, Bowness, Conishead Priory, Flookborough, Milnthorpe, etc. Ascends to 300 yards.

46. *Glaucium luteum*, Scop. (Yellow Horned Poppy). Native. Maritime. English type. Range 1. Sands of the seashore.

C. Nethertown, St. Bees, Seascale, and Coulderton.—(J. Robson, Rev. F. Addison, W. Hodgson, Whitehaven Cat.)

L. On the seashore near Bardsea and Winder Hall.—(Aiton.) Furness shore at Bardsea, and west shore of Walney Island.—(Miss Hodgson.) On the coast at Flookborough.—(Otley.) Shore west of Humphrey Head.—(C. J. Ashfield, B.) Cartmel sands and Roosebeck.—(Woodward.) Walney Island.—(Atkinson.)

ORDER FUMARIACEÆ.

48. *Corydalis claviculata*, DC. (Climbing or White Fumitory). Native. British type. Range 1. Woods and heathy places. Frequent.

C. Near Wythburn.—(Rev. A. Ley.) On Gowbarrow Fells, near Collier Hag.—(W. Hodgson.) Hallin Fell, Ullswater. —(Rev. J. E. Leefe.) Common in Ennerdale.—(Whitehaven Cat.) Ashness Gill and Barrow Woods, near Keswick.— (Winch, Watson, B.) Near Dalegarth, in Eskdale.—(J. Robson.) Llanthwaite Woods, Crummock.—(W. B. Waterfall.)

W. Thornthwaite (Haweswater), foot of Long Sleddale, thatched roofs in Kentmere and at the foot of Isan Parle's Cave.—(Lawson.) Spital Wood and side of a ditch on the east side of Kendal Castle.—(Wilson, Gough.) Place Fell, near the Patterdale slate quarries.—(W. Hodgson.) Little Langdale Tarn and other places round Windermere.—(W. Foggitt, Miss Edmonds.) Easedale.—(Miss Beever.) Nab Scar, Rydal.—(J. C. Melvill.) Helm Crag and Buthar Crags, Grasmere.—(A. W. Bennett.) In the Troutbeck Valley, below the uppermost inn with a verse from Shenstone on the sign.—(B.) Drawn from Fox Ghyll in Miss Wilson's collection.

L. Among stones near the mines and on the fells, Coniston. —(Miss Beever.) Dry stony places on Furness Fells.—(Atkinson.) Long Scar near Holker Hall.—(Aiton.) Rowdsey Wood and Bankend Wood, near the Duddon.—(Miss Hodgson.)

Corydalis lutea, DC. (Yellow Fumitory). Alien. An occasional straggler from gardens.

C. Roadside, near the Lorton Yew-tree, with *Meconopsis*. —(B.) Stock Bridge and Little Mill, Egremont.—(Whitehaven Cat.) Walls about homesteads, Aspatria.—(W. Hodgson.)

W. Walls at Eamont Bridge.—(B.)

Corydalis solida, DC. (Solid-rooted Fumitory). Alien. An occasional straggler from garden cultivation.

C. Vale of Newlands and Vicar's Island, Derwentwater.—(J. B. Davies, L.) Foot of Wastwater.—(J. Robson.)

W. Watsfield, Kendal, near a garden, and Leven's Park, near Milnthorpe.—(Gough.)

L. Ulverstone.—(E. Robson.)

50. *Fumaria capreolata*, L. (Ramping Fumitory). Colonist. British type. Range 1. Cultivated fields. Much the commonest species of the genus in the Lake country. It is recorded both by Lawson and Hudson. Ascends to 300 yards over Penrith. Of the sub-species, *Borœi* and *confusa* are frequent; *pallidiflora* rare; *muralis* I have not seen in the Lake district.

51. *Fumaria officinalis*, L. (Common Fumitory). Colonist. British type. Range 1. Cultivated ground; not seen in the heart of the Lakes, about Keswick, Ambleside, or Coniston, but abundant in places on the outskirts, as at Penrith, Greystoke, and Burton in Kendal. Very abundant in light sandy ground about Aspatria.—(W. Hodgson.)

52. *Fumaria micrantha*, Lag. (Small-flowered Fumitory). Colonist. English type. Range 1. Gathered in September 1849 by Rev. F. J. A. Hort near Ambleside and Hawkshead. —(Botanical Gazette, ii. 54.)

ORDER CRUCIFERÆ.

55. *Cakile maritima*, Scop. (Purple Sea-Rocket). Native. Maritime. British type. Range 1. Sands of the seashore. Rare.

C. Whitehaven, Seascale, Parton, Coulderton Point, St.

Bees.—(J. Robson, W. Foggitt, Rev. F. Addison.) Also northwards as far as Silloth.—(W. Hodgson.)

L. Walney Island and Roosebeck in Furness. First recorded by Lawson in 1688. Not seen about Flookborough, Grange, or Arnside.

56. *Crambe maritima*, L. (Sea-Kale). Native. English type. Maritime. Range 1. Sands of the seashore. Rare.

C. Coast between Ravenglass and Bootle. (Mr. Wood, in Bot. Guide.) Between Maryport and Flimby.—(Harriman.) Now extinct.—(W. Hodgson.) Coulderton Point.—(J. Robson.)

L. Roosebeck in Furness.—(Woodward, in Bot. Guide.) West shore of Walney Island at Summerhill, and further south; collected by the country people, probably for pickling. —(T. Gough, C. Bailey, Miss Hodgson.)

Coronopus didyma, Sm. (Lesser Wart-Cress). Alien.

C. Waste ground at Whitehaven.—(W. Foggitt.)

58. *Coronopus Ruellii*, Gaertn. (Common Wart-Cress). Native. English type. Range 1. Waste ground. Very rare.

C. Seaton near Workington.—(Mr. Tweddle.) Probably now extinct.—(W. Hodgson.) Edge of highway, Solway shore at Dubmill. Very rare.—(W. Hodgson.)

W. On the waste near Kent Terrace, Kendal.—(T. Gough.)

60. *Thlaspi arvense*, L. (Penny Cress). Colonist. British type. Range 1. Cultivated ground. Very rare.

C. St. Bees.—(Whitehaven Cat.)

62. *Thlaspi alpestre*, L. (Perfoliate Shepherd's Purse, or Small Thorow Cress). Native. Highland type. Range 2-3. Rocky places. Rare.

C. Believed to have been gathered at a moderate elevation under the steep end of Skiddaw.—(Watson.)

W. Precipices of east face of Helvellyn, just beneath the summit.—(Rev. A. Ley!, 1871 and 1881.) By the road between Kendal and Ambleside.—(M.) With *Teesdalia* on rocks in Fusedale.—(W. Foggitt.) The Fusedale plant is the var. *occitanum*. Ray and Hudson record *T. perfoliatum* as occurring in most pastures in Westmoreland and Cumberland. In Hudson's second edition the name is altered to *alpestre*, which occurs in several stations in the adjacent parts of Northumberland, Durham, and Yorkshire.

63. *Capsella Bursa-pastoris*, Gaertn. (Common Shepherd's Purse). Native. British. Range 1-2. Roadsides and waste places. Common; ascending to 350 yards (Watson); to 300 yards at Shap and over Penrith (B.).

65. *Teesdalia nudicaulis*, R., Br. (Naked-stalked Teesdalia). Native. British type. Range 1. Rocky and sandy places. Not infrequent.

C. Banks and hills round Derwentwater.—(Watson.) Rocks by the roadside at Grange and High Lodore.—(W. Foggitt, J. Britten, R. Holland.) Raven Crag, Thirlmere.—(M.) Vale of St. John.—(Miss Edmonds.) Walls by Thirlmere, extending nearly a mile.—(W. Foggitt.) Thief Gill, Dean, near Cockermouth.—(M.) Mockerkin, and at Bowness in Ennerdale.—(Whitehaven Cat.) Priest Crag and Gowbarrow Fells, in some places very plentiful. Also in Swineside, at the foot of Carrock Fell, and in a lane near Hanging-Shaw Moss. —(W. Hodgson.)

W. By Common Holme Bridge, near Clibburn.—(Lawson.) Dry hillsides round Ullswater.—(W. Hodgson.) Hallen Fell, Ullswater.—(Leefe.) Fusedale, Ullswater, abundant.—(W. Foggitt.) Goat Scar, Long Sleddale; Applethwaite, by the road to Ings.—(T. Gough.) Kentmere Scars.—(F. C. Roper.)

Iberis amara, L. (Bitter Candytuft). Alien. An occasional straggler from gardens.

C. East Mill, Aspatria, a straggler.—(W. Hodgson.)
W. Waste near Kent Terrace, Kendal.—(T. Gough.)
L. By Jacklands Tarn, Low Furness.—(Miss Hodgson.)

Lepidium Draba, L. (Whitlow Pepperwort). Alien. Casually in waste ground.

W. Whitbarrow.—(L.)
L. A patch by the Ulverstone road at Newby Bridge, 1853.—(Borrer.)

69. *Lepidium Smithii*, Hook. (Hairy Pepperwort). Native. British type. Range 1. Roadsides; frequent. Maryport, Keswick, Lodore, Rosthwaite, Clibburn, Grasmere, Windermere, Hawkshead, Yewdale, Coniston village, Newby Bridge, Newton, Colton Beck Bridge, Cartmel, etc.

70. *Lepidium campestre*, L. (Field Pepperwort). Native. British type. Range 1. Cultivated fields and roadsides. Less frequent than *L. Smithii*. Barrow, Cleator, etc.

72 a. *Cochlearia officinalis*, L. (Common Scurvy-Grass). Native. British type. Range 1-3. The type confined to the coast.

C. Rocky shore at St. Bees Head.—(Rev. F. Addison, W. Hodgson.)
W. Shore at Arnside.—(C. Bailey.)
L. On the shore at Flookborough.—(B.)

Var. *alpina*. Rocks by the river in Lowther Park.—(Rev. A. Ley.)

C. Abundant in the sykes of Great Gable over Styhead Pass and down Kirk Fell into Mosedale (500-700 yards).—(Watson, B.) Scaw Fell.—(J. Robson.) Black Sail Pass.—

(Britten and Holland.) Rosset Ghyll and Piers Ghyll.—(J. C. Melvill.) Ennerdale, from Gillerthwaite upwards to Great Gable.—(W. Hodgson.) On Rydal Fell.—(F. C. Roper.)

W. Place Fell, Hallin Fell, and plentiful in the Ullswater streamlets up to Hayes Water and the head of Kirkstone Pass. First recorded by Fardon and Woods, Striding-edge, Glenridding, Grisedale Tarn, and other places round Helvellyn, ascending to 700 yards.—(Watson, W. Foggitt, B., W. Hodgson.) Mardale, Rossgill, Swindale, and Long Sleddale. —(Wilson, Watson, etc.) Abundant on the Troutbeck end of High Street.—(J. C. Melvill, B.)

L. Mountain above Coniston Lake.—(Woodward.) Seathwaite Fells, and carried down thence to the Duddon estuary.—(Miss Hodgson.) Ghylls on Dobby Shaw.—(Miss Hodgson.)

72*. *Cochlearia danica*, L. (Danish Scurvy-Grass). Native. Maritime. British type. Range 1. Waste ground along the coast.

C. On the shore between Bootle and Ravenglass.—(Mr. Wood, in Bot. Guide.) Coulderton and Nethertown.—(Whitehaven Cat.)

L. Walney Island. First recorded by Lawson in 1688. Summerhill Bank, Walney Island.—(Miss Hodgson.) Barrow, in the excavations for the new docks, 1867.—(C. Bailey.) Humphrey Head.—(Dr. Windsor.) The inland stations which are recorded for this in the Botanist's Guide and elsewhere belong to *C. alpina.*

72. *Cochlearia anglica*, L. (English Scurvy-Grass). Native. Maritime. British type. Range 1. Like the last, confined to the seashore.

C. On the coast near Whitehaven.—(W. Foggitt.) St. Bees Head.—(J. Robson.) Workington shore.—(W. Dickinson.)

L. Walney Island.—(C. Bailey.) Tidal banks, Low

Meathop.—(C. Bailey.) Near Flookborough and Conishead Priory.—(Aiton.) (What I saw there was *C. officinalis.*) Grange-over-Sands.—(W. Matthews.)

Armoracia rusticana, Baumg. (Horse-Radish). Alien. An occasional straggler from cultivation.

W. Ambleside, and waste ground near Ferry Inn, Windermere.—(W. Foggitt.) Lake Head, Windermere, 1882.—(B.)

L. Grange-over-Sands.—(B.)

74. *Subularia aquatica,* L. (Awlwort). Native. Highland type. Range 1-3. Lakes and tarns. Rare.

C. Ennerdale Lake.—(J. Robson.) Floutern Tarn, between Loweswater and Ennerdale, 1250 feet.—(D. Oliver!) Borrowdale and Derwentwater.—(Rev. Wm. Hind.) Bowscale Tarn. —(Rev. R. Wood.)

W. Red Tarn, Helvellyn (800 yards).—(G. C. Druce.)

77. *Draba incana,* L. (Twisted-podded Whitlow Grass). Native. Xerophilous. Highland type. Range 1. Limestone cliffs. Very rare in the Lake district, but several stations known in the east of Westmoreland and Cumberland and the adjacent parts of Yorkshire.

C. Wastdale.—(J. Robson.) (Needs confirmation, as it is a limestone species.) Figured in Rooke's Flora, marked 'Seascale, May 1871.'—(W. Hodgson.)

W. Force Beck Fall near Shap (300 yards).—(Watson!)

Draba muralis, L. (Speedwell-leaved Whitlow Grass). Mentioned by Hudson as a Westmoreland plant, but I know of no station within the limits of the county, though there are several in the adjacent parts of Yorkshire.

79. *Draba verna,* L. (Common Whitlow Grass). Native. British type. Range 1. Frequent on walls and roofs in the lower zone.

Var. *brachycarpa*. On wall-tops in High Furness, and from the shore to the top of Birkrigg near Ulverstone.—(Miss Hodgson!)

Camelina sativa, Crantz (Common Gold of Pleasure). Alien. An occasional weed of cultivated ground.

C. Workington Mill field.—(Mr. Tweddle.) Thackthwaite, near a stream flowing into Dacre Beck, at the east end of Ullswater, 1878.—(W. Hodgson.) A garden weed at Ghyllbank College, Whitehaven.—(W. Hodgson.)

Koniga maritima, R., Br. (Sweet Alyssum). Alien. An occasional straggler from garden cultivation.

L. Grange-over-Sands.—(Hindson.)

84. *Cardamine amara*, L. (Bitter Ladies' Smock, or Large-flowered Bitter-Cress). Native. British type. Range 1. Stream-sides and damp places.

C. Bearpot and Moorside Woods near Lamplugh.—(Mr. Tweddle, W. Hodgson.) Powbeck and banks of the Ehen and its tributaries.—(Whitehaven Cat.) Fitz near Aspatria.—(Rev. J. Dodd.) Shore of Derwentwater near Lodore.—(Watson.) Moist woods round Ullswater and in Ennerdale.—(W. Hodgson.)

W. Beck Mills near Kendal.—(T. Gough.) Windermere and Grasmere, rare.—(F. Clowes.) Ambleside, in plenty.—(C. Bailey.) Drawn from Fox Ghyll in Miss Wilson's collection.

L. Leven banks, Newby Bridge.—(W. Foggitt.) Once found at Coniston.—(Miss Beever.) Bogs in Furness and Cartmell.—(Aiton.) Brook near Falls farm, north-west of Ulverstone.—(Miss Hodgson.)

85. *Cardamine pratensis*, L. (Meadow Ladies' Smock, or Cuckoo Flower; local name, 'May Flower'). Native. British

type. Range 1-3. Common in damp grassy places, ascending to Red Tarn and the high springs of Helvellyn, 900 yards, and Coniston Old Man.—(B.) *Flore pleno* recorded by Lawson, from Little Strickland pasture; and found also by Mr. Hodgson at Aspatria.

86. *Cardamine hirsuta*, L. (Hairy Ladies' Smock, or Bitter-Cress). Native. British type. Range 1-2. Walls and damp places, ascending to 550 yards on Great Gable (B.); to 600 yards (Watson).

86 *b. Cardamine sylvatica*, Link. (Wood Ladies' Smock, or Bitter-Cress). Native. British type. Range 1-2. Woods and damp rocks, frequent, ascending to the top of Kirkstone Pass, 500 yards.—(B.)

87. *Cardamine impatiens*, L. (Impatient Ladies' Smock, or Bitter-Cress). Native. Xerophilous. Intermediate type. Range 1. Woods in the limestone tract.

C. In Ennerdale at Bowness, and walls by the roadside near Barrow Falls.—(R. Holland.)

W. Shap Abbey and in several places near Kendal.—(Hudson, Wilson, Gough.) Witherslack Woods, at the foot of Whitbarrow.—(C. Bailey.) Abundant in Middlebarrow Wood, Arnside.—(J. C. Melvill, B.)

88. *Arabis thaliana*, L. (Common Wall-Cress). Native. British type. Range 1. Walls and dry rocks. Frequent. Ascends to the summit of Castle Crag in Borrowdale, 300 yards.—(B.)

89. *Arabis petræa*, Lam. (Alpine Rock-Cress). Native. Highland type. Range 2. High slate rocks. Very rare.

C. Ravine of the Wastwater Screes, 600 feet in perpendicular height.—(Mr. Wood, in Bot. Guide.) Scawfell.—(J. Robson.)

Arabis stricta, Huds. (Bristol Rock-Cress).

C. Lamplugh Hall and Pardshaw Hall near Loweswater.
—(M.) No doubt a mistake; probably *A. hirsuta* intended, which Mr. Hodgson has gathered there.

92. *Arabis hirsuta*, R., Br. (Hairy Rock-Cress). Native. Xerophilous. British type. Range 1-2.

C. Lamplugh Hall, in Lowestar, and about Red Hills Limeworks, Penrith.—(W. Hodgson.) Ullock and Pardshaw Crag.—(W. B. Waterfall.)

W. Frequent on the limestone scars about Penrith, Shap, and Kirkby Lonsdale, and from Kendal by way of Whitbarrow and Huttonroof Crags to Arnside. Ascends to 500 yards.— (Watson.) Hills between Keswick and Thirlmere, 1000 feet.

L. Rocks at Plumpton; on the beach at Bardsea and in Rowdsey Wood.—(Miss Hodgson.) Not included in Whitehaven Cat., or in my lists made at Keswick and Ambleside.

94. *Turritis glabra*, L. (Long-podded or Smooth Tower-Mustard). Native. English type. Range 1. Dry banks. Very rare.

C. Stainburn near Workington.—(Mr. Tweddle!)

W. In the red sandstone tract at Clibburn near Penrith.— (Lawson.)

95. *Barbarea vulgaris*, R., Br. (Yellow Rocket). Native. British type. Range 1. Frequent by stream-sides in the low country, ascending into Great Langdale, to Dacre Beck, over Ullswater, and to 300 yards at Shap.

Barbarea præcox, R., Br. (Winter-Cress). Alien. An occasional straggler from gardens.

W. A few plants in the road between Shap and Shap Abbey.—(Watson.)

L. Included in Miss Hodgson's catalogue for Furness; no station cited.

ORDER CRUCIFERÆ. 37

98. *Nasturtium officinale*, R., Br. (Water-Cress). Native. British type. Range 1. Frequent in ditches and streams, ascending to 250 yards.—(Watson.) I have seen it as high near Shap and at Rossgill, in both cases associated with *Myosotis palustris*, and mixed with *Epilobium alsinifolium* at the foot of Great Dodd in the Vale of St. John.

99. *Nasturtium terrestre*, R., Br. (Rocket). Native. British type. Range 1. Damp places. Rare.

C. In the 'Meadows' near Wigton.—(Prof. Oliver!) Edges of pools about Aspatria and Gilcrux.—(W. Hodgson.)

W. Near the mill-dam at Kirkby Lonsdale.—(Hindson.)

100. *Nasturtium sylvestre*, R. Br. Native? English type. Range 1. Damp places. Very rare.

L. Barrow in Furness.—(W. Foggitt.)

102. *Sisymbrium officinale*, Scop. (Common Hedge-Mustard). Native. British type. Range 1. Common by roadsides and in waste ground, ascending in Borrowdale to Stonethwaite (B.); 250 yards (Watson); as high at Coniston, and to 300 yards near Penrith Beacon.

Sisymbrium Irio, L., is reported in the Whitehaven Catalogue from the banks of the Marron, but I have seen no specimen from the Lake district.

Sisymbrium Sophia, L., has occurred as a garden weed at Carlton Hill, Penrith.—(W. Hodgson.)

107. *Erysimum Alliaria*, L. (Garlic Treacle-Mustard, or Jack by the Hedge). Native. British type. Range 1. Common in woods and on hedge-banks, ascending to 300 yards.—(Watson.)

Cheiranthus Cheiri, L. (Wallflower). Alien. Old walls. Rare.

C. Carlisle Castle.—(Winch.) Penrith Castle.—(C. Bailey.) Scaleby Castle.—(M.) Millom Castle.—(Whitehaven Cat.) Dacre Castle.—(W. Hodgson.)

W. Coast cliff between Silverdale and Arnside Point.—(C. J. Ashfield, J. C. Melvill.)

Hesperis matronalis, L. (Scentless Dame's Violet). Alien. An occasional straggler from gardens. Recorded by Mr. Nicolson in the Dillenian edition of Ray's Synopsis from Dalehead, on the west of Thirlmere and Grasmere.

C. By the river Ellen, from Cockbridge downwards.—(W. Hodgson.)

W. Side of the Troutbeck road near Bowness.—(F. C. Roper.)

Brassica oleracea, L. (Wild Cabbage). Alien. Recorded from near Arnside Point by my relative, E. Robson, in the old Botanist's Guide. I have no later information about it as a plant of the district, and could not find it there in 1883.

114. *Brassica polymorpha,* Syme. Colonist. English type. Range 1. Cultivated fields and waste places. Frequent. Ascends to 250 yards in Troutbeck Valley, and 300 yards at Shap.

116. *Sinapis arvensis,* L. (Charlock, or Wild Mustard; local name, 'Field Kale.') Colonist. British type. Range 1. Frequent in cultivated fields, ascending to 300 yards near Penrith Beacon. It is perhaps on the whole the greatest pest in the way of weeds the Lakeland farmers have to contend with.

117. *Sinapis alba,* L. (White Mustard). Colonist. English type. Range 1.

C. Reported from St. Bees in the Whitehaven Catalogue.

W. Seen as a weed amongst potatoes at Holme Mill near Milnthorpe.—(B.)

118. *Sinapis nigra*, L. (Black, or Common, Mustard). Colonist. English type. Range 1.

C. Once seen as a weed in a field of turnips near Dalehead, on the west side of Thirlmere.—(B.)

120. *Sinapis tenuifolia*, R., Br. Denizen. English type. Range 1.

C. Walls of Penrith Castle.—(Balfour.) (Not seen there in 1883.) Walls and draw-dikes of Carlisle Castle.—(Winch.) (Originally published as *S. muralis.*) A weed on gravel walks in the garden at Whitefield House near Overwater.—(W. Hodgson.)

122. *Sinapis monensis*, Bab. Native. Maritime. Atlantic type. Range 1.

C. In many places along the Cumberland coast-line. First recorded by Lawson. Northwards as far as Silloth banks, where it is extremely abundant.—(W. Hodgson.) Allonby, Flimby, and plentiful about Sellafield and Seascale.

L. Walney Island, and thence along the shore of the mainland to Grange; also first recorded by Lawson.

123. *Raphanus Raphanistrum*, L. (Wild Radish, or Jointed Charlock). Colonist. British type. Range 1. Frequent in cultivated fields. Ascends to 300 yards.—(Watson.) I have seen it at that height over Coniston. Both the yellow- and white-flowered varieties occur.

Rapistrum rugosum, All. Alien.
C. A weed amongst flax at Penrith, 1883.—(B.)

Neslia paniculata, Desv. Alien.
L. Waste ground at Grange-over-Sands.—(F. C. Roper.)

ORDER RESEDACEÆ.

125. *Reseda Luteola*, L. (Dyer's Weed, Yellow-weed, or Weld). Native. British type. Range 1. Dry banks. Rare.

C. In several places about Workington and Cockermouth.—(M.) Railway banks, Harrington, Workington, and Brigham.—(Whitehaven Cat.) Brigham limeworks, Cockermouth.—(W. Hodgson.)

W. Abundant in limestone quarries at Kendal.—(T. Gough.) Banks of the Lune at Kirkby Lonsdale.—(Hindson.)

L. Waste ground at Newland.—(Miss Hodgson.) Roadside at Kents Bank, on limestone.—(B.)

126. *Reseda lutea*, L. (Wild Mignonette, or Base Rocket). Denizen. English type. Range 1.

C. Railway slope near Coulderton, 1876; not truly wild.—(W. Hodgson).

L. Waste ground near Ulverstone; doubtfully wild.—(Miss M. A. Ashburner.)

ORDER CISTACEÆ.

128. *Helianthemum vulgare*, Gaertn. (Rock Rose). Native. Xerophilous. British type. Range 1-2. Limestone cliffs and banks. Frequent.

C. Limestone rocks, Slapestones Brow near Penrith.—(W. Hodgson.) Clints at Isell.—(Whitehaven Cat.)

W. Common along the limestone from Lowther, Shap and Kendal, by way of Whitbarrow, to Arnside and Milnthorpe, but I did not see it on Farleton Knot or Huttonroof Crags. Ascends to 360 yards.—(Watson.) A variety with red-tipped petals, and the whole plant smaller on Whitbarrow.—(J. C. Melvill.)

L. Bardsea Park, Humphrey Head, Rowdsey Wood, Hill

above Grange, etc. Aiton records a var. *tomentosum* from Humphrey Head.

130. *Helianthemum canum*, Dun. (Hoary Dwarf-rock Rose). Native. Xerophilous. Intermediate type. Range 1. Limestone cliffs. Rare.

W. Limestone cliffs between Kendal and Arnside; Scout Scar, Cunswick Scar, Witherslack, and Whitbarrow. First recorded by Lawson.

L. Plentiful at the end of Humphrey Head, a station recorded by Ray. I saw it there in plenty in 1883.

ORDER VIOLACEÆ.

132. *Viola palustris*, L. (Marsh Violet). Native. British type. Range 1-3. Common from the lake-sides up to the highest springs. Noted at 800 yards on Helvellyn, and 900 yards on Scawfell Pike.—(B.)

133. *Viola odorata*, L. (Sweet Violet). Denizen. English type. Range 1. Woods and hedge-banks; I believe a native in the limestone tract, but often introduced.

C. Hedge-banks at Cockermouth and Brackenthwaite.—(B.) Roadside near Watermillock House, and summit of Slapestones near Penrith.—(W. Hodgson.) Chapel House, Linethwaite, and other places near Whitehaven. A white variety near Sandwith.—(Whitehaven Cat.)

W. Not uncommon at Kirkby Lonsdale.—(Hindson.) Hedge-banks at Watsfield, and other places near Kendal.—(T. Gough.) Crosby Ravensworth.—(Watson.) Hedge-bank above Townend, in the Winster Valley.—(B.) Hedge-banks at Clawthorpe and Burton in Lonsdale.—(B.)

L. About Ulverstone and Holker Hall.—(Aiton.) Several places near Ulverstone; flowers white, lilac or white.—(Miss Hodgson.) Hedge-banks at Grange and Allithwaite.—(B.)

134. *Viola hirta*, L. (Hairy Violet). Native. Xerophilous. English type. Range 1. Woods on the limestone. Rare.

C. Near the entrance to Dalemain Park from Penrith.—(B., W. Hodgson.)

W. Frequent about Kirkby Lonsdale.—(Hindson.) Barrowfield and other woods near Kendal.—(Gough, Watson.) Ascends to 300 yards.—(Watson.) Middlebarrow Wood, Arnside.—(J. C. Melvill.) Lowther Woods, and quarries near Great Strickland.—(B.) Whitbarrow and Farleton Knot, ascending to the limestone pavement of the summit.—(B.)

L. Waitham Wood near Holker Hall, and woods round Conishead Priory.—(Aiton.) Plumpton Woods, along with *V. odorata*.—(Miss Hodgson.) Yewbarrow, and banks between Grange and Lindale.—(B.)

135. *Viola sylvatica*, Fries. (Dog Violet). Native. British type. Range 1-3. Grassy places. Common, ascending to 550 yards on Great Gable, (B.); 850 yards on Striding-edge (B.); and 850 yards on Grisedale Pike (Watson).

V. Reichenbachiana, Bor., was gathered by Mr. Watson near Keswick, and by Miss Hodgson in Lake Lancashire, about Newfield and Seathwaite, and at 700 feet on the banks of Cockley Beck, and by Rev. H. Higgins and myself near Grange-over-Sands.

135*. *Viola canina*, L. (Dog Violet). (*V. flavicornis*, Smith.) This was cultivated by Mr. Watson at Thames Ditton from a plant sent by Professor Oliver from Warwick Bridge in Cumberland. In Mr. Watson's list for Shap both *canina* and *flavicornis* are marked. Miss Hodgson includes true *canina* in her list for Lake Lancashire, without mention of any special station, and Mr. W. Hodgson gives *flavicornis* as growing on the north shore of Ullswater at Gowbarrow.

W. Typical. Islands of Ullswater.—(W. B. Waterfall.)

136. *Viola tricolor*, L. (Field Pansy, or Heart's-ease). Native. British type. Range 1. Roadsides and cultivated fields, both the type and var. *arvensis*. Ascends to Watendlath (Winch); 320 yards (Watson).

136*. *Viola lutea*, Huds. (Yellow Mountain Pansy). Native. Scottish type. Range 1-2. Mountain pastures. Frequent.

C. Hills round Keswick, especially Skiddaw, in the pasture crossed just after leaving Lalrigg.—(Watson, B.) In the Cocker Valley, on Mellfell, and in Ennerdale, and near Hassness in the Buttermere Valley.—(Winch, B.) Dent Hill near Whitehaven.—(Rev. F. Addison.) Brigham near Cockermouth.—(M.) Penruddock, Ennerdale.—(Whitehaven Cat.) Abundant round Ullswater, especially about Dockray.—(W. Hodgson.) Drawn from St. John's Vale in Miss Wilson's collection, and also gathered there by Mr. W. Matthews.

W. Rydal Head and Fairfield.—(J. C. Melvill.) Swindale Moor, Haweswater.—(B.) No record for Lake Lancashire.

Var. *Curtisii*, Forst.
L. Sands of the shore in Walney Island.—(Rev. A. Ley.)

ORDER DROSERACEÆ.

138. *Drosera rotundifolia*, L. (Round-leaved Sundew). Native. British type. Range 1-3. Frequent in bogs, ascending to 500 yards on the Styhead Pass and round Hayes Water (B.); 650 yards (Watson). Much the commonest of the three species.

139. *Drosera intermedia*, Hayne (Dwarf Sundew). Native. English type. Range 1. Bogs; only at a low elevation.

C. Borrowdale and Ullock Moss near Portinscale.—(Mr. Tweddle.) Ulpha.—(J. Robson.) North shore of Wastwater.—(Rev. A. Ley.) Foot of Carrock Fell near Stone Ends farm.—(Rev. R. Wood.) Figured in Rooke's MS. Flora, where it is marked 'Ennerdale.'—(W. Hodgson.) Wedholm Flow near Wigton.—(W. B. Waterfall.)

W. Witherslack, Brigstear Moss, and Foulshaw Moss near Kendal. First recorded by Lawson and Wilson. Gilpin Bridge near Bowness.—(J. Woods.) Not uncommon about Kirkby Lonsdale.—(Hindson.)

L. Pools near Coniston Tarns.—(Miss Beever.) Abundant on Plumpton and other low-lying moss-ditches.—(Miss Hodgson.) Roundsea Mosses.—(Aiton.)

140. *Drosera anglica*, Huds. (Great Sundew). Native. Scottish type. Range 1.

C. Ullock Moss near Portinscale.—(W. Dickinson.) Helvellyn.—(J. Flintoft.) Moss at Grange, abundant.—(J. C. Melvill.) Seathwaite in Borrowdale.—(Miss Edmunds.) Side of Crummock.—(W. B. Waterfall.)

W. Foulshaw Moss and Brigstear Moss near Kendal. First recorded by Wilson.

L. Stickle Pike, Donnerdale.—(W. F. Miller.)

ORDER POLYGALACEÆ.

141. *Polygala vulgaris*, L. (Milkwort). Native. British type. Range 1-2. Common in grassy places; ascending to 1700 feet on Saddleback, and to 600 yards on Grisedale Pike and Helvellyn.—(Watson.) The plant of the higher levels is all *P. depressa*.—(Wender.) I have seen var. *oxyptera* on Whitbarrow.

ORDER CARYOPHYLLACEÆ.

146. *Dianthus Armeria*, L. (Deptford Pink). Native. English type. Range 1.

W. At Orton near Great Strickland.—(Lawson.) No recent confirmation for the district, but reported from Nunnery by Mr. Cooke.

ORDER CARYOPHYLLACEÆ.

Dianthus cæsius, L. (Mountain Pink).

L. Very rare. On the limestone rocks in Furness.—(Aiton.) (Doubtless a mistake.)

150. *Dianthus deltoides*, L. (Maiden Pink). Native. English type. Range 1.

C. Foot of Skiddaw.—(Mr. Cooke.)

W. Sandy hill below Common, Holme Bridge near Great Strickland.—(Lawson.) I searched for it there without success in 1883.

L. Common pastures in High Furness.—(Aiton.)

Saponaria Vaccaria, L. Alien.

C. A weed in garden ground at Ghyllbank College, Whitehaven.—(W. Hodgson.)

151. *Saponaria officinalis*, L. (Soapwort). Denizen. English type. Range 1. Hedge-banks and river-sides. Rare.

C. Derwent-side at Workington.—(Mr. Tweddle.) Hedge in Aspatria village (Rev. J. Dodd); now extinct (W. Hodgson). At Torpenhow, and also at Oughterside in the same neighbourhood; not strictly wild.—(W. Hodgson.) Santon Bridge.—(Whitehaven Cat.) Keswick.—(L.)

W. Force Bridge and Hawes Bridge near Kendal.—(T. Gough.) Formerly at Akebeck near Pooley, and near Howtown.—(W. Hodgson.) Kirkby Lonsdale, frequent on the banks of the Lune.—(Hindson.)

L. A few plants near Conishead Priory.—(Aiton.)

152. *Silene inflata*, Smith (Bladder Campion, or Catchfly). Native. British type. Range 1. Road-sides and dry banks. Frequent. Ascends to 300 yards.—(Watson.)

153. *Silene maritima*, With. (Sea Campion, or Catchfly). Native. British. Range 1-2.

C. Along the coast at Workington, Harrington, Whitehaven, St. Bees, and Ravenglass.—(T. J. Foggitt, Rev. F.

Addison, etc.) Coledale Pass, under Grasmoor.—(W. B. Waterfall.) In Borrowdale by the stream at Stockley Bridge, and lower down about Seatoller and Grange, below Castle Crag, the shores of Derwentwater and on the Catbells, about 500 yards, and in the vale of Newlands.—(Watson, B.) Near Brackenthwaite in the vale of Lorton.—(M.) Hills south of Ennerdale.—(Rev. F. Addison.) Piers Ghyll.—(J. C. Melvill.)

W. Arnside.—(Bailey, B.) All along the shore; cliffs at the head of Deepdale.—(Rev. A. Ley.) In Dungeon Ghyll above the upper waterfall, 500 yards.—(B.)

L. Furness shores, at Biggar Bank, Bardsea, Flookborough, Humphrey Head, and Grange, abundant.—(Miss Hodgson, W. Foggitt, C. Bailey, etc.) Inland on rocks in several places. Near the summit of Coniston Old Man.—(Miss Beever.)

Silene nutans, L. (Nottingham Catchfly).

C. Moorland Close, and Dean near Workington.—(M., W. Dickinson.) Doubtless a mistake. Is *S. anglica* intended?

159. *Silene acaulis*, L. (Cushion Pink, or Moss Campion). Native. Highland type. Range 2. Damp mountain rocks. Rare.

C. Crags of Mickledore, and on the black rocks of Great End, 1500-2000 feet.—(Watson, J. Robson.)

W. Helvellyn, St. Sunday's Crag, Grisedale, Deepdale Crags, and by Tongue Ghyll waterfall.—(Woods, Winch, etc.) Rocks near Grisedale Tarn.—(Flintoft, W. Hodgson.) Above Rydal Mount, where it was shown to me in 1847 by Wordsworth.—(J. Sidebotham.) Rocks in Langdale in several places.—(J. Sidebotham.)

160. *Lychnis alpina*, L. (Alpine Campion). Native. Highland type. Range 3.

C. Hobcarten Fell near Brackenthwaite, in a damp narrow

ravine at an elevation of about 2000 feet. First recorded by R. Matthews, in Phytologist, ii. 185. Since gathered by Borrer, Oliver, Dickinson, etc. Gathered by Rev. A. Ley (on Hobcarten Crag) in 1881.

L. Coniston Old Man.—(R. Potter, in Report of Botanical Record Club, 1879, p. 52.)

162. *Lychnis Flos-cuculi*, L. (Ragged Robin). Native. British type. Range 1-2. Frequent in damp grassy places. Ascends to 500 yards at Hayes Water.—(B.)

163. *Lychnis diurna*, Sibth. (Red Campion). Native. British type. Range 1. Frequent in woods and on shaded rocks, ascending to 250 yards in Borrowdale, and as high in the Troutbeck Valley and about Haweswater. Local name, 'Head-aches,' in West Cumberland.

164. *Lychnis vespertina*, Sibth. (White Campion). Native. British type. Range 1. Hedge-banks and forage fields. Not seen about Ambleside, Keswick, or Coniston, but only on the outskirts of the Lake district, as about Whitehaven, Grange, and Penrith, where it ascends to 300 yards. I have gathered a pink-flowered variety in Meathop Moss. Not infrequent in West Cumberland, from Skiddaw towards the coast.—(W. Hodgson.) Gathered by Mr. F. C. Roper at Winster over Windermere. Local name, 'Thunder-flower,' in West Cumberland.

165. *Lychnis Githago*, Lam. (Corn Lychnis, or Corn Cockle). Colonist. British type. Range 1. Frequent in cultivated fields.

166. *Moenchia erecta*, Smith (Upright Pearlwort). Native. English type. Range 1. Dry banks. Very rare.

C. St. Bees and Coulderton.—(Whitehaven Cat.)

167. *Sagina apetala*, L. (Annual Small-flowered Pearlwort). Native. English type. Range 1.

C. Roadsides between Penrith and Carleton, and not uncommon round Ullswater.—(W. Hodgson.) Foot of walls between Eamont Bridge and Penrith, and in the village streets at Greystoke.—(B.)

W. Walls at Ambleside, especially near the old church.—(J. Sidebotham.)

L. Coniston, a troublesome weed in garden walks.—(Miss Beever.) Foot of walls at Grange-over-Sands.—(B.)

For *S. maritima*, which is sure to be found along the coast, I have no record, and I have searched for it in vain about Flookborough, Grange, and Arnside.

167*. *Sagina ciliata*, Fries. Native. English type. Range 1.

L. Crevices of limestone walls at Kents Bank.—(B.) It is given as common in the Whitehaven Catalogue, but I fear confusion with *S. apetala*, as the latter is not mentioned.

168. *Sagina procumbens*, L. (Procumbent Pearlwort). Native. British type. Range 1-3. Walls and grassy places. Frequent. I have seen it at 600 yards in the springs on Great Gable, and Mr. Watson up to 710 yards.

171. *Sagina nodosa*, E. Meyer (Knotted Spurrey). Native. British type. Range 1-2. Sands of the sea-shore and inland springs.

C. Lily Hall near Workington.—(Mr. Tweddle.) Shore at St. Bees.—(J. Robson.) Limestone quarry at Blencow Station. —(B.) Hanging Shaw Moss, New Cooper, Aspatria.—(W. Hodgson.)

W. Troutbeck Holm near Great Strickland.—(Lawson.) Shap Fell and Kendal Fell up to 550 yards.—(Watson.) Salt marsh between Arnside and Milnthorpe.—(B.) Swamps on Brantfell above Bowness, and near the summit of Whitbarrow.—(B.) Moor Divock, Sharrow Bay, and Sandwyke, Ullswater.—(W. Hodgson.)

L. Fells above Grange-over-Sands and on the shore.—(T. J. Foggitt, B.) Railway embankment between Cark and Ulverstone.—(B.)

172. *Spergula arvensis*, L. (Corn Spurrey). Native. British type. Range 1. Cultivated fields. Common. Ascends to 300 yards.—(Watson.) I have seen it as high over Coniston and near Penrith Beacon. A terrible pest to the Abbey Holme farmers, who call it 'Dodder.'—(W. Hodgson.)

173. *Honkeneya peploides*, Ehrh. (Sea Purslane). Native. Maritime. British type. Range 1. Sands of the coast-line.

C. Parton, Coulderton, St. Bees, and Seascale.—(W. Foggitt, Rev. F. Addison, Whitehaven Cat., W. Hodgson.)

L. Plentiful on Barrow Island.—(C. Bailey.) Isle of Walney and all round the Furness shore.—(Miss Hodgson.) Shore marshes at Cark and Flookborough.—(B.)

174. *Spergularia marginata*, Syme (Sea Sandwort). Native. Maritime. British type. Range 1. Salt marshes.

C. On the Solway shore at Dubmill.—(W. Hodgson.)

W. Shore near the railway at Arnside.—(C. Bailey.)

L. On the shore west of Humphrey Head, and about Cark and Flookborough.—(Dr. Windsor, B.) School Bank, Isle of Walney, Morecambe shores at Greenodd, Tridley Marsh near Ulverstone.—(Miss Hodgson.) Grange-over-Sands.—(W. Matthews.)

174*. *Spergularia neglecta*, Syme (Sea Red Sandwort). Native. British type. Range 1.

C. Cliffs south of Whitehaven.—(Rev. F. Addison.) (This may not unlikely prove to be *S. rupestris*.) Saltcoats.—(Rev. R. Wood.)

L. Shore at Bardsea.—(Aiton.) With the last in the salt marshes at Cark and Flookborough.—(B.)

D

174*. *Spergularia rubra*, Fenzl. (Red Sandwort). Native. British type. Range 1. Sandy soil. Rare.

C. Hensingham.—(Whitehaven Cat.) Foot of walls round Penrith Beacon, 250-300 yards.—(B.)

W. Sandy ground on Common Holme Bridge, above Clibburn.—(B.) Quarry near Clibburn railway station.—(W. Hodgson.)

L. Cartmel, in fields near the sea.—(Aiton.) (The other species probably intended here.)

178. *Arenaria serpyllifolia*, L. (Thyme-leaved Sandwort). Native. British type. Range 1-2. Walls and dry rocks. Frequent. Ascends to 300 yards on the limestone cliffs of Shap Common (B.); to 500 yards (Watson).

Var. *leptoclados*. On the limestone of Humphrey Head.— (Miss Hodgson !)

180. *Arenaria verna*, L. (Vernal Sandwort). Native. Scottish type. Range 1-3. Rocky and grassy places amongst the hills; not so common in the Lakes as in the lead districts of Durham and Yorkshire.

No record for Cumberland.

W. Fairfield, and rocks above Red Tarn, Helvellyn, up to 800 yards.—(W. Foggitt, B.) Bleawater, at the head of Mardale.—(Watson.) Limestone hills south of Kendal. First recorded by Hudson as *Arenaria saxatilis*, and afterwards as *A. laricifolia*. Roadside near Arnside Tower.—(B.) Farleton Knot, and hill between Witherslack and the Winster Valley. —(B.)

L. With double flowers on Hampsfield Fell, Cartmel, 400 feet.—(Miss Hodgson.) Limestone banks between Grange and Lindale.—(B.)

182. *Arenaria trinervis*, L. (Plaintain-leaved Chickweed or Sandwort). Native. British type. Range 1. Hedge-banks and thickets. Frequent.

184. *Stellaria nemorum*, L. (Wood Stitchwort). Native. Scottish type. Range 1-2. Damp woods. Not infrequent.

C. Near Aspatria Mill.—(Rev. J. Dodd.) Burdoswald and Moorside Hall near Lamplugh.—(M.) Penruddock.— (Whitehaven Cat.) Woods above Lodore and Grange in Borrowdale.—(C. Bailey.) Fairly plentiful round Ullswater; many brooklets at Watermillock.—(W. Hodgson.)

W. Damp woods round Windermere.—(F. Clowes.) High up in Fusedale, above Howtown.—(W. Hodgson.) Laverock Lane and Thornley Hill near Kendal.—(Hudson, Wilson, Gough.) By Casterton Mill near Kirkby Lonsdale.—(Sir J. E. Smith.) Stock Ghyll Force.—(Rev. J. H. Thompson.) I have no record for Lake Lancashire.

185. *Stellaria media*, With. (Common Chickweed). Native. British type. Range 1-2. Everywhere common in waste ground. Ascends to 520 yards.—(Watson.) I have traced it up to 500 yards at Hayes Water, Kirkstone Pass, and on Coniston Old Man.

Var. *neglecta* is a common Lakeland variety.

186. *Stellaria Holostea*, L. (Greater Stitchwort). Native. British type. Range 1. Hedge-banks and thickets. Common. Ascends to 300 yards.—(Watson.) At 250 yards in Borrowdale above Lodore, and in the Troutbeck Valley.—(B.)

187. *Stellaria glauca*, With. (Glaucous Marsh Stitchwort). Native. English type. Range 1. Damp grassy places. Rare.

C. Muncaster Woods, Ravenglass.—(J. Robson.) Linethwaite near Whitehaven.—(Whitehaven Cat.)

188. *Stellaria graminea*, L. (Lesser Stitchwort). Native. British type. Range 1-2. Common by stream-sides, and in grassy swamps. Ascends to 480 yards.—(Watson.) Round Hayes Water, 500 yards.—(B.)

189. *Stellaria uliginosa*, Murr. (Bog Stitchwort). Native. British type. Range 1-3. Swamps at all levels. Frequent. Ascends to the high springs of High Street, Coniston Old Man, Helvellyn, Great Gable, and Scawfell Pike, 700 yards. —(B.) Mostly associated with *Montia fontana* and *Chrysosplenium oppositifolium*.

192. *Cerastium glomeratum*, Thuill. (Clustered Chickweed). Native. British type. Range 1-2. Roadsides and waste ground. Frequent. Ascends to 350 yards.—(Watson.) To 400 yards in Hag Ghyll, Troutbeck.

193. *Cerastium triviale*, Link. (Mouse-ear). Native. British type. Range 1-4. Everywhere common in grassy places. I have a note of it at 900 yards on Helvellyn, and Watson at 1010 yards on the same mountain.

194. *Cerastium semidecandrum*, L. (Little Mouse-ear Chickweed). Native. British type. Range 1.

C. Common in the neighbourhood of Whitehaven.—(Whitehaven Cat.)

194* *Cerastium tetrandrum*, Curt. (Tetrandrous Mouse-ear Chickweed.) Native. Maritime. British type. Range 1. Walls and dry banks on the coast.

C. Allonby.—(Rev. R. Wood.)

L. North-end rabbit warren, Isle of Walney, and wall-tops near Ulverstone.—(Miss Hodgson.)

195. *Cerastium arvense*, L. (Field Chickweed). Native. English type. Range 1.

C. Amongst the red sandstone quarries of Penrith Beacon, 250-300 yards.—(B.)

196. *Cerastium alpinum*, L. (Alpine Mouse-ear). Native. Highland type. Range 3.

W. Sparingly on the Striding-edge Crags, Helvellyn, at about 900 yards. First recorded by Woods.—(B.) North side of Fairfield.—(F. Clowes.) Deepdale Crags, Fairfield.—(J. C. Melvill.) Langdale.—(J. Sidebotham.)

Claytonia perfoliata, Don. Alien. Waste ground.

C. Between Coulderton and the shore, Rose Hill near Seascale, Calder Bridge, and other places.—(Whitehaven Cat.) Dub Beck, near Mill-yeat, Whitehaven; planted forty years ago some three miles higher up the stream by Mr. Dickinson, and spreading downwards.—(W. Hodgson.)

Claytonia alsinoides, Sims. An occasional straggler from gardens.

L. Hedge-bank at the south end of the village of Sawrey.—(B.)

ORDER LINACEÆ.

Linum usitatissimum, L. (Common Flax). Alien. Cultivated about Penrith and Clibburn.

W. Not uncommon in corn-fields about Arnside. It appears to have been at some time cultivated in that neighbourhood.—(J. Sidebotham.) A weed in a potato-field at Clibburn.—(B.)

L. Newland and Horrace farm, Ulverstone.—(Miss Hodgson.)

200. *Linum perenne*, L. (Perennial Blue Flax). Native? Germanic type. Range 1.

W. At Crosby Ravensworth, and between Shap and Threaplands.—(Lawson.) Modern confirmation wanted. A bare elevated limestone country does not seem a very likely locality for this species.

202. *Linum catharticum*, L. (Purging Flax). Native. British type. Range 1-2. Dry banks and grassy places. Frequent. Noted at 500 yards at Hayes Water. It is specially common in the limestone tract, and ascends to the top of Whitbarrow, Farleton Knot, and Huttonroof Crags, 550 yards on Great Gable (B.); 600 yards (Watson).

203. *Radiola millegrana*, Sm. (All-seed.) Native. British type. Range 1. Sandy moors. Rare.

C. Swinside near Keswick.—(Black's Guide.) Mouth of the Ehen near Sellafield.—(M. Chambers.) Plentiful on the north shore of Wastwater; coast sand-hills at Drigg.—(Rev. A. Ley.) Springs at the south base of Dent Hill near Egremont.—(W. Hodgson.) Long Meg.—(Mr. Cooke.)

W. Clifton and Clibburn Moors near Penrith.—(T. Lawson.) Formerly on Foulshaw Moss near Milnthorpe.—(T. Gough.)

ORDER MALVACEÆ.

204. *Malva moschata*, L. (Musk Mallow). Native. British type. Range 1. Dry banks. Not infrequent, especially on limestone.

C. Common in Cumberland.—(Winch.) Eskdale roadside.—(J. Robson.) Eaglesfield.—(Whitehaven Cat.) Ullswater; frequent about quarries, Birkcrag Quarry, etc.—(W. Hodgson.) Redhill Quarry near Penrith.—(B.)

W. Common about Kendal.—(T. Gough.) Frequent round Kirkby Lonsdale.—(Hindson.) Ferry Inn, Sawrey, Newby Bridge, and other places round Windermere.—(F. Clowes, W. Foggitt, etc.)

L. Bardsea beach and Bankside, Cartmel.—(Aiton.) Coniston village.—(B.) Allithwaite and Humphrey Head.—(C. J. Ashfield.) Cartmel, Haverthwaite, Ulverstone, etc.—(Miss Hodgson.) Banks at Grange and Cark.—(B.)

205. *Malva sylvestris*, L. (Common Mallow). Denizen. British type. Range 1. Waste ground; always near villages and farm-houses. Whitehaven, Watermillock, Kendal, Tebay, Clibburn, Clawthorpe, Hackthorpe, Bowness, Ulverstone, Allithwaite, etc. Often grown in cottage gardens.

206. *Malva rotundifolia*, L. (Dwarf Mallow). Denizen. British type. Range 1. In similar places to the last. Very rare.

C. Cockermouth.—(Whitehaven Cat.)

L. Plentiful in a farm-yard near the railway, Kents Bank. —(C. Bailey.) Allithwaite near Cartmel.—(C. J. Ashfield.) Near a cottage at the north end of the village of Cark.—(B.)

Althæa officinalis, L. (Marsh Mallow).

L. Near Bardsea and Broughton in Furness.—(Aiton.) I suspect a mistake in identification; *Malva sylvestris* perhaps intended.

ORDER TILIACEÆ.

211. *Tilia parvifolia*, Ehrh. (Lime, or Linden Tree). Denizen. English type. Range 1.

C. An old battered tree in the rock near the Borrowdale bowder-stone.—(Watson.)

W. On the shore cliffs at New Barns near Arnside.—(B.)

L. Exposed limestone rocks of Humphrey Head, both on the west and east faces.—(C. Bailey, B.)

Tilia intermedia, DC., is frequently to be seen in parks and hedgerows, and *T. grandifolia*, Ehrh., occasionally, as at Portinscale and Victoria Bay, on the west side of Derwentwater, and the foot of Gummers How, Windermere.—(C. Bailey.) *T. argentea* is planted by the side of the road at Portinscale.—(C. Bailey.)

ORDER HYPERICACEÆ.

214. *Hypericum Androsæmum*, L. (Tutsan). Native. Atlantic type. Range 1. Damp thickets. Not infrequent.

C. By the Borrowdale stream between Seathwaite and Seatollar.—(B.) Snebra, Barrowmouth, Wormghyll, and Eskdale.—(Whitehaven Cat.) Amongst the coast cliffs near St. Bees Head.—(Rev. F. Addison.) Barrow Force, Keswick.—(J. Otley.)

W. Stock Ghyll, Ferry Woods, Lady Holme Island, and other places round Windermere. First recorded by Lawson. Barrowfield Wood, and other places round Kendal.—(T. Gough.) Whitbarrow Woods.—(W. Foggitt.) Banks of Rydal Lake.—(Balfour.) Not infrequent in Rydal Park.—(J. C. Melvill.)

L. Woods at Ulverstone, and foot of Coniston Water at Lake Bank.—(Miss Hodgson.) Near Kirkby and Dalton in Furness.—(Aiton.)

Hypericum elatum, Ait. Alien.

W. Side of the road between Bowness and Windermere Ferry.—(F. C. Roper.)

215. *Hypericum perforatum*, L. (St. John's Wort). Native. British type. Range 1. Hedge-banks and thickets. Common. Ascends from the shore at Flookborough to 250 yards in the Troutbeck Valley.

216. *Hypericum dubium*, Leers (Imperforate St. John's Wort). Native. English type. Range 1. In similar places to the last, but much less frequent.

C. North side of Ullswater.—(Balfour.) St. Bees.—(Whitehaven Cat.)

W. Hills south-west of Kendal.—(Watson.)

L. Near Coniston Lake, not frequent.—(Miss Beever.) Isle of Walney and lanes near Penny Bridge.—(Miss Hodgson.) Bushy places, Humphrey Head.—(Dr. F. A. Lees.)

217. *Hypericum quadrangulum*, L. (Square-stalked St. John's Wort). Native. British type. Range 1. Grassy swamps, frequent. Ascends to 300 yards.—(Watson.)

218. *Hypericum humifusum*, L. (Trailing St. John's Wort). Native. British type. Range 1. Dry banks and woodland hedgerows. Frequent.

220. *Hypericum pulchrum*, L. (Slender St. John's Wort). Native. British type. Range 1-2. Shaded rocks and borders of heaths. Frequent. Ascends to 500 yards in Langdale, 550 yards on Great Gable (B.); 600 yards (Watson).

221. *Hypericum hirsutum*, L. (Hairy St. John's Wort). Native. British type. Range 1. Woods and hedge-banks, especially on limestone.

C. Camerton and Clifton near Workington.—(M.) Northwest side of Ullswater.—(Balfour.) Dalemain Park.—(B.)

W. Wood near Celleron.—(W. Hodgson.) Whitbarrow and other woods, on the limestone about Kendal, ascending to 300 yards.—(Gough, Watson, etc.) Woods at Lowther, Great Strickland, Clibburn, and by the stream near Brougham Hall.—(B.)

L. Middlebarrow Wood, Arnside.—(J. C. Melvill.) Hedgebanks between Grange and Lindale.—(B.) Shore of Windermere near Ferry Inn.—(B.) Roadside between Grange and Cartmel.—(B.)

222. *Hypericum montanum*, L. (Mountain St. John's Wort). Native. Xerophilous. English type. Range 1. Woods and thickets on the limestone. Not infrequent.

C. Coldfell near Egremont.—(J. Robson.) Requires confirmation.

W. Rocks by the rivulet between Anna Well and Shap.—(Lawson, Watson.) Scout Scar, Cunswick Scar, Whitbarrow, Witherslack, and other limestone hills between Kendal and Arnside. First recorded by Lawson. Lowther Woods.—(B.) Near the shore at New Barns, Arnside.—(B.)

L. Woods between Grange and Lindale, with *H. hirsutum*.—(B.) Humphrey Head.—(C. Bailey.) Rocks in Bardsea Park and Hagg Wood near Holker.—(Aiton.) Near Cartmel Well.—(Mr. Jackson.)

223. *Hypericum elodes*, L. (Marsh St. John's Wort). Native. Atlantic type. Range 1. Peaty bogs. Not infrequent.

C. Braystones Tarn, Wormghyll, Nethertown Tarn, and other places near Whitehaven.—(Whitehaven Cat., J. Robson.) Wastdale.—(Rev. A. Ley.) Dent Hill.—(Rev. F. Addison.) Ullock Moss near Portinscale.—(Otley.) Birker Moss and Aitchar Moor.—(W. Dickinson.)

W. Underbarrow Common, Kendal.—(T. Gough.)

L. Near Rampside in Furness.—(Miss Beever.) Bogs in the Isle of Walney.—(Miss Hodgson.)

Hypericum calycinum, L. (Large-flowered St. John's Wort). Alien. Parks and roadsides. Introduced.

C. Ennerdale, a garden escape.—(Whitehaven Cat.) Irton Woods near Ravenglass.—(J. Robson.)

W. Roadsides near Brathay in several places.—(J. Sidebotham.)

ORDER ACERACEÆ.

225. *Acer campestre*, L. (Common Maple). Native. English type. Range 1. Woods and hedges. Rare.

C. Plantations at Waterfoot, Ullswater.—(W. Hodgson.)

W. Kirkby Lonsdale, frequent in woods.—(Hindson.) Hedge

near Meathop.—(B.) Road to Silver Howe from Grasmere. —(F. C. Roper.)

L. Lane and woods near Furness Abbey.—(C. Bailey, B.) Hedges between Lindale and Grange-over-Sands.—(B.)

Acer pseudo-platanus, L. (Sycamore). Alien. Common in plantations and about farm-houses up to 500 yards. It is one of the commonest trees planted to shelter the scattered farm-houses. The finest specimens I have seen are at Greystoke, Furness Abbey, and at the bottom of Glenridding. I doubt its being a true native, but it is often self-sown, as in the crevices of the limestone pavement of Farleton Knot and Huttonroof Crags.

Staphylea pinnata, L. (Bladder-nut Tree). Alien. An occasional stray from gardens.

W. Roadside near Rydal.—(Balfour.)

L. Reported from Finsthwaite Woods near Newby Bridge. —(Borrer.) A fine tree in the grounds of Furness Abbey Hotel.—(J. C. Melvill.)

ORDER GERANIACEÆ.

Erodium maritimum, Sm. (Sea-Stork's Bill).

C. On the coast at St. Bees.—(M.) Not confirmed in the recent Whitehaven list, and I suspect *E. cicutarium*, which grows there, may have been mistaken for it.

228. *Erodium cicutarium*, Sm. (Hemlock Stork's Bill). Native. British type. Range 1. Common on the sandy seashore all along the coast. Inland I have seen it abundantly in sandy ground at Clibburn, and Wilson reports it from the brow of Kendal Fell, and Mr. Hodgson from the village green at Dalston near Carlisle.

Var. *pilosum* (Boreau). On the sands of the Isle of Walney. —(Miss Hodgson.)

Erodium moschatum, Sm. (Musky Stork's Bill). Alien? 'In pratis siccis in comitatu Westmorlandico.'—(Hudson.) Upland pastures in High Furness.—(Aiton.) Occasional plants near Ambleside by roadsides.—(J. Sidebotham.)

Geranium phæum, L. (Dusky Crane's Bill). Alien. An occasional garden escape.

C. Roadside at Dockray.—(Mrs. King.) North side of Ullswater near a cottage.—(Balfour.) In Nether Wastdale at Strands.—(J. Robson.) Near some cottages at Wastdale Head.—(C. Bailey.) Lamplugh, St. Bees, Mockerkin, and Prior Scale.—(Whitehaven Cat.) Hedge-banks at Stocks Nook, Watermillock.—(W. Hodgson.) Pardshaw Hall near Lorton.—(W. B. Waterfall.)

W. Banks of the Brathay near Ambleside.—(C. Bailey.) Lane near Kirkby Lonsdale.—(Leefe.) Biggins near Kirkby Lonsdale.—(Hindson.) Skirts of a wood on both sides of the road between Burnside and Kendal.—(T. Gough.)

L. Ulverstone, and under trees in Little Croft Park.—(Miss Hodgson.)

Geranium nodosum, L. (Knotty Crane's Bill). Alien. Like the last, but less frequent.

C. With *G. phæum*, on the north side of Ullswater.—(Balfour.) Not seen by Mr. W. Hodgson, who thinks the Floshgate *striatum* may have been meant. Reported by Mr. Wright as gathered near Thirlmere.—(Borrer, Phytologist, ii. 430.)

W. Kirfit Hall, Casterton.—(Hindson.)

Geranium striatum, L. Alien. A casual straggler from gardens.

C. Reported by Mr. Wright as gathered near Flimby.—

(Woods, in Comp. Bot. Mag. i. 296.) North side of Ullswater, at Floshgate, where Mr. W. Hodgson showed me it in 1883.

230. *Geranium sylvaticum*, L. (Wood Crane's Bill). Native. Scottish type. Range 1-2. Frequent in meadows in the heart of the Lake country about Derwentwater, Ullswater down to Pooley Bridge, Thirlmere, Grasmere, Borrowdale, Watendlath Valley, Buttermere, Ambleside, Coniston Water, etc.

C. Eskdale, Wythop, and other places about Whitehaven. —(Whitehaven Cat.)

W. Mardale and Shap.—(Watson.) Ascends to 560 yards. Oxenholme and Staveley.—(B.) Frequent about Kirkby Lonsdale.—(Hindson.)

L. By Seathwaite Tarn Beck at Newfield.—(Miss Hodgson.)

231. *Geranium pratense*, L. (Great Crane's Bill). Native. British type. Range 1-2. Frequent in meadows, and by the lake-sides, ascending from the Furness salt marshes at Plumpton to 350 yards in Mardale.

232. *Geranium pyrenaicum*, L. (Mountain Crane's Bill). Denizen. English type. Range 1.

C. Dale Head on the west side of Thirlmere.—(Black's Guide.) Yearton Hall near Beckermet.—(L.) Requires confirmation. In Martineau's Guide this last station is ascribed to *G. rotundifolium*.

234. *Geranium pusillum*, L. (Small-flowered Crane's Bill). Native. English type. Range 1. Dry grassy places. Rare.

C. Not uncommon about Whitehaven.—(Whitehaven Cat.) Etterby Scar near Carlisle.—(M.) Sandy field at Penrith.— (B.)

W. Windermere.—(L.) Sandy ground at Clibburn.—(W. Hodgson, B.)

235. *Geranium molle*, L. (Common Crane's Bill). Native. British type. Range 1-2. Grassy places. Frequent. Ascends to the foot of Honister Crag (Britten and Holland); to top of Castle Crag in Borrowdale (B.); the limestone cliffs of Shap common; to 500 yards (Watson).

236. *Geranium dissectum*, L. (Cut-leaved Crane's Bill). Native. British type. Range 1. Roadsides and forage fields. Frequent. Ascends to 300 yards.—(Watson.)

237. *Geranium columbinum*, L. (Dove's Foot Crane's Bill). Native. English type. Range 1. Roadsides and grassy places, especially on the limestone.

C. Cockermouth and St. Bees.—(Whitehaven Cat.) On the shore at Seascale.—(J. Robson.) Roadside near Coulderton hamlet, sparingly.—(W. Hodgson.)

W. In several places on the limestone about Kendal. First recorded by Wilson. Plentiful about Arnside Knot.—(C. Bailey, B.)

L. Ascending the hill from Coniston to Hawkshead.—(T. J. Foggitt.) Fell foot near Newby Bridge.—(L.) Roadside at Newton.—(B.) Plumpton woods and Furness shore at Bardsea.—(Miss Hodgson.) Kents Bank, Humphrey Head, and woods between Grange and Lindale.—(B.)

238. *Geranium lucidum*, L. (Shining-leaved Crane's Bill). Native. British type. Range 1. Roadsides and rocky places. Not infrequent.

C. Brigham, Pardshaw, Cockermouth, and near Gill-foot, Egremont.—(Whitehaven Cat.) Vale of St. John, Ormathwaite, Lodore, Grange, and up Borrowdale to the top of Castle Crag, 300 yards. First recorded by Winch. Common round Ullswater and about Penrith; also about Gilcrux and Tallentire. —(W. Hodgson, B.) Patterdale.—(Winch.)

W. Rydal, Troutbeck, and other places about Windermere.

—(M.) Askham.—(T. J. Foggitt.) Common about Shap and Lowther.—(Watson, B.) Amongst the limestone hills between Kendal and Arnside, common. — (Watson, B.) Arnside.—(B.) Kirkby Lonsdale, Burton in Lonsdale, and Huttonroof Crags.—(Hindson, B.) A brilliant ornament to the romantic dales of Westmoreland.—(Sir J. E. Smith.)

L. Hedge-banks near Cartmel.—(C. J. Ashfield.) Plentiful by roadsides at Sawrey, Hawkshead, and Newton.—(B.) West-end lane, Ulverstone.—(Miss Hodgson.) Borwick Lodge; not seen near Coniston.—(Miss Beever.) Common about Allithwaite and Grange.—(B.)

239. *Geranium Robertianum*, L. (Herb Robert). Native. British type. Range 1-2. Woods and hedge-banks. Common. Ascends to 400 yards in Great Langdale, and 500 yards on Coniston Old Man. A white variety at Torver, and Fox How near Ambleside (Miss Beever); on Huttonroof Crags (B.); and at Kendal in the lane to Jenkin Crag (T. Gough).

240. *Geranium sanguineum*, L. (Red Crane's Bill). Native. British type. Range 1. Frequent along the coast-line, amongst the sand-hills, and on the cliffs. Allonby, Maryport, Egremont, Seascale, Walney Island, Humphrey Head; inland on Scout Scar, Whitbarrow, and other limestone hills between Kendal and Arnside, ascending to 300 yards.— (Watson.) Shap.—(Watson.)

The Walney Island *G. lancastriense*, Withering, originally described by Ray and figured by Dillenius (Hortus Elthamensis, p. 163, tab. 136, fig. 163), was first gathered by Lawson, who writes, 'Thousands hereof I have found on the Isle of Walney, and have sent roots to Edinburgh, York, London, and Oxford, where they keep their distinction.' Extends from Summerhill, its northern limit, to the south end of the Biggar bank, a full mile; both in the beach gravels and on the grassy sward.—(Miss Hodgson.)

ORDER BALSAMINACEÆ.

242. *Impatiens Noli-me-tangere*, L. (Wild Balsam, or Touch-me-not; local name, 'Old Woman's Purse'). Denizen. Local type. Range 1.

C. Scale Hill, over Crummock.—(Woods.) A little east of Keswick along the Penrith road, near the stream that runs past a garden higher up.—(B.) Duddon Bridge.—(J. Robson.)

W. Banks of the streams about Rydal and Ambleside, in several places, especially in Scandale and Stock Ghyll. First recorded by Lawson. Ghyll near Whittington Hall.—(Hindson.)

L. Near Coniston Water.—(P. J. Woodward.) Foot of the Nite, near Yew-tree, and near the railway station, Coniston.—(Miss Beever.) In a small gully at Coniston which is passed on the ascent of the Old Man.—(J. C. Melvill.) Roadside near Storrs Hall, south of Bowness.—(B.)

ORDER OXALIDACEÆ.

243. *Oxalis Acetosella*, L. (Wood Sorrel; local name, 'Cuckoo's Bread and Cheese'). Native. British type. Range 1-4. Woods and shaded rocks. Common. Ascends to 850 yards on Saddleback (Watson); 900 yards on Helvellyn (B.); and 1010 yards on Scawfell Pike (Watson). A variety with pink flowers seen both at Ambleside and Coniston (Miss Beever), and in the wood at Dunmallet at the foot of Ullswater Lake (W. Hodgson).

ORDER CELASTRACEÆ.

245. *Euonymus europæus*, L. (Spindle Tree). Native. English type. Range 1. Woods, especially on the limestone.

C. In the Derwentwater woods at Barrow and Lodore. First recorded by Winch. In Rooke's manuscript Flora is a drawing of this species marked 'Gowbarrow Park, 1851.'—(W. Hodgson.)

W. In the lane to Fowl Ing, Kendal.—(T. Gough.) Middlebarrow Wood, Arnside.—(B.) Kirkby Lonsdale, not uncommon.—(Hindson.) Woods near Witherslack Hall.—(B.)

L. Windermere shore near the Ferry, and about Newby Bridge.—(B.) Woods of Yewbarrow and between Grange and Lindale.—(B.) Woods at Plumpton, Haverthwaite, Bardsea, and elsewhere in Furness.—(Miss Hodgson.)

ORDER RHAMNACEÆ.

246. *Rhamnus catharticus*, L. (Buckthorn). Native. English type. Range 1. Woods, especially on the limestone.

C. Slapestones How, Penrith, only one bush. — (W. Hodgson.)

W. Hedges near Great Strickland.—(T. Lawson.) Frequent about Kirkby Lonsdale.—(Hindson.) Cunswick and other limestone woods south of Kendal.—(Gough, Watson.) Hedges at Clawthorpe.—(B.) Common about Arnside.—(B.)

L. Islands and shores of Windermere.—(F. Clowes, etc.) Woods between Grange and Lindale.—(B.)

247. *Rhamnus Frangula*, L. (Black Alder). Native. English type. Range 1. Woods and peat-mosses in several places.

C. Ullock, Cockshot, Cass and Lodore woods near Keswick.—(Watson, B.) Lamplugh.—(Rev. F. Addison.) Pardshaw near Lorton.—(W. B. Waterfall.) Thornthwaite.—(W. Dickinson.)

W. Thorny Holme, Whinfell Forest. — (T. Lawson.) Cunswick Wood, Kendal.—(T. Gough.) About Rydal Water.—(J. Otley.) Meathop Moss and woods south of Witherslack Hall.—(B.) Middlebarrow Wood, Arnside.—(B.)

L. Islands of Windermere.—(F. Clowes.) Brathay Woods, Colton Beck Wood, and Stribers peat-bog near Cartmel.—(Miss Hodgson.) Humphrey Head.—(Dr. Windsor.)

ORDER LEGUMINIFERÆ.

248. *Sarothamnus scoparius*, Koch. (Broom.) Native. British type. Range 1-2. Woods and thickets, ascending to nearly 500 yards on the Catbells, west of Derwentwater.—(Watson.)

249. *Ulex europæus*, L. (Furze, Whin, Gorse). Native. British type. Range 1. Hillsides. Universally distributed through the lower zone, of which it is one of the most characteristic and conspicuous plants. Ascends above 300 yards on Latrigg. A variety with double flowers noted at Clappersgate near Ambleside, by C. Bailey.

250. *Ulex Gallii*, Planch (Autumnal Furze). Native. English type. Range 1. Universally distributed through the district. Common in the Crummock, Ennerdale, and Wastwater valleys, down to Gosforth, Lamplugh, Cockermouth, and Whitehaven; Ullswater, Penrith Beacon (where it ascends to 300 yards), Threlkeld, Blencow, Clibburn, Temple Sowerby, Haweswater, Ambleside, Bowness, Coniston, Troutbeck, Newton, Grange, Ulverstone, etc. It must have covered a considerable portion of the ancient forest of Inglewood.—(W. Hodgson.)

251. *Genista tinctoria*, L. (Dyer's Weed). Native. English type. Range 1. Heaths. Frequent, ascending to 250 yards over Haweswater.—(Watson.)

Genista pilosa, L.

L. Frequent on rocks in High Furness.—(Aiton.) Not

seen by any one else, and I have little doubt *G. tinctoria* intended.

253. *Genista anglica*, L. (Small Whin). Native. British type. Range 1. Heathy places. Rare.

C. Moors at Bootle and Drigg.—(J. Robson.) Church Moss, Beckermet.—(Whitehaven Cat.) Seaton Moor near Workington.—(W. Hodgson.) Wigton.—(W. B. Waterfall.)

W. Kendal.—(J. Sidebotham.) Edge of Clibburn Moss.— (W. Hodgson.)

L. Parks at High Furness and Cartmel.—(Aiton.) On the right bank of the Leven, a mile below Newby Bridge.— (J. Sidebotham.)

254. *Ononis arvensis*, L. (Rest-harrow). Native. British type. Range 1. Coast sand-hills and dry inland pastures. Frequent. Ascending to 300 yards near Shap. A spinose variety occurs on the Furness shore at Roosebeck and Flookborough. This is probably the plant given by Aiton as *O. spinosa*. I have not seen the true *spinosa* within our limits.

257. *Anthyllis vulneraria*, L. (Lady's Finger). Native. British type. Range 1. Dry pastures, especially in the limestone tract.

C. Abundant on the railway banks at St. Bees, etc.— (Whitehaven Cat.) Maryport and Nethertown.—(J. Robson.) Near Dearham Bridge, Maryport; also in Whinbarrow Quarries, Aspatria.—(W. Hodgson.)

W. Sandy fields at Clibburn.—(B.) Abundant about Kendal.—(T. Gough.) Shap.—(Watson.) 320 yards. Limestone quarries at Winder near Askham.—(W. Hodgson.)

L. Humphrey Head and woods between Grange and Lindale.—(B.) Hampsfield Fell, Leybarrow Crags, and in the park at Dalton in Furness.—(Miss Hodgson.)

Medicago sativa, L. (Lucerne). Alien. Cultivated fields. Rare.

C. Fields near Whitehaven.—(Watson.) Etterby Scar.—(W. Duckworth.)

W. Railway embankment at Carnforth.—(J. C. Melvill.)

260. *Medicago lupulina*, L. (Black Medick). Native. British type. Range 1. Dry grassy places. Frequent. Ascending to 300 yards on Shap Common. 320 yards.—(Watson.)

264. *Melilotus officinalis*, Willd. (Yellow Melilot). Colonist. English type. Range 1. Roadsides and forage fields. Rare.

C. Near Cross-side, Egremont, and Whitrigg Station.—(Whitehaven Cat.) Workington and Etterby.—(Dickinson.) Langanby.—(T. Lawson.) Railway cutting on the Derwent branch, near Bullgill Station.—(W. Hodgson.)

W. Near Foulshaw House.—(Wilson.)

L. Abundant on the railway bank at Grange.—(T. Gough, Miss Hodgson.)

Melilotus arvensis, Wallr. Alien.

W. On the railway embankment at Arnside.—(C. Bailey.)

L. Railway banks east of Grange, on both sides of the county boundary between Westmoreland and Lancashire.—(W. Matthews!)

Melilotus parviflora, Lam. Alien.

C. North shore of Ullswater at Floshgate.—(W. Hodgson!)

L. A solitary plant near the farmstead at Bowstead Gates near Ulverstone.—(Miss Hodgson!) (Wrongly given as *M. vulgaris* in Journ. Bot.)

266. *Trigonella ornithopodioides*, DC. Native. English type. Range 1.

C. Workington Warren.—(Mr. Tweddle.) Confirmation

wanted. Workington Warren has pretty well disappeared: now under tillage, or covered with iron furnaces, collieries, etc.—(W. Hodgson.)

267. *Trifolium repens*, L. (Dutch Clover, White Clover). Native. British type. Range 1-3. Grassy places. Common. Ascending to 500 yards on Walna Scar.—(Miss Hodgson.) 550 yards on Great Gable.—(B.) 700 yards.—(Watson.)

271. *Trifolium pratense*, L. (Red Clover). Native. British type. Range 1-2. Grassy places. Common. Ascending to 500 yards at Hayes Water and Kirkstone Pass.

272. *Trifolium medium*, L. Native. British type. Range 1. Thickets and hedge-banks throughout the lower zone. Frequent. Ascending to 300 yards.—(Watson.)

Trifolium hybridum, L. (Alsyke Clover). Alien. Is now common in cultivated fields throughout the lower zone.

274. *Trifolium maritimum*, Huds. Native. English type. Maritime. Range 1.

C. On the shore at Braystones.—(J. Robson.) The nearly-allied Mediterranean *T. supinum*, L., was found by Mr. W. Hodgson, with the other casuals, at Floshgate, Ullswater.

275. *Trifolium arvense*, L. (Hare's-foot Clover). Native. British type. Range 1. Sandy ground. Rare.

C. Flimby near Workington.—(W. Dickinson.) Hodbarrow, Braystones.—(Whitehaven Cat.) St. Bees.—(W. B. Waterfall.) Railway station, Maryport.—(W. Hodgson.)

W. Sandy field between Clibburn village and the railway station, abundant.—(B.)

L. Border of a field at Cark.—(Dr. F. A. Lees.)

277. *Trifolium striatum*, L. Native. English type. Range 1. Sandy ground. Very rare.

C. Shore at St. Bees.—(M. Chambers, W. Dickinson.) Miscopied *T. strictum* in Linton's Guide.

L. Near Grange-over-Sands.—(C. Bailey.)

280. *Trifolium fragiferum*, L. (Strawberry Clover). Native. English type. Range 1. Sandy soil, especially near the sea.

W. Banks of the river Gilpin between Gilpin's Bridge and Raven's Lodge.—(C. Bailey.) Above the Bridge Inn at Levens.—(T. Gough.)

L. Sand-hills near Barrow in Furness.—(W. Foggitt.)

Trifolium resupinatum, L. Alien.

C. Floshgate, Ullswater; introduced with foreign corn, 1882, along with *Melilotus parviflora*, *Centaurea melitensis*, and other aliens.—(W. Hodgson.)

281. *Trifolium procumbens*, L. (Hop Clover). Native. British type. Range 1. Dry grassy places. Frequent; ascending to the limestone cliffs of Shap Common, 300 yards.

Trifolium agrarium, L. Alien.

W. Abundant in a forage field ascending the hill west of Witherslack Hall, 1883.—(B.)

L. Clover-fields at Plumpton near Ulverstone.—(Miss Hodgson.)

282. *Trifolium minus*, Relh. Native. British type. Range 1-2. Dry grassy places. Frequent; ascending to 250 yards on Brantsfell, and in Troutbeck Valley; 300 yards on the limestone of Shap Common, and in the red sandstone quarries of Penrith Beacon; 350 yards.—(Watson.) In several lists *T. filiforme* is included, but I have not seen Lakeland examples of the true plant.

283. *Lotus corniculatus*, L. (Bird's-foot Trefoil). Native. British type. Range 1-2. Everywhere common in grassy places; ascending to 500 yards on Catbells, and Skiddaw to 600 yards.—(Watson.)

Var. *tenuis* in fields at Whitehaven.—(Watson.) A villose variety at Sandy Gap and Biggar Marsh, Isle of Walney.—(Miss Hodgson.) A var. with fleshy leaves on the shore at Flookborough.

Lotus angustissimus, L. Grew some ten years ago in a railway cutting on the Derwent branch near Bullgill Station. —(W. Hodgson.) It is also recorded from Hysemoor and Clifton, on the authority of the late Mr. W. Dickinson, but the naming needs confirmation.

284. *Lotus major*, Scop. Native. British type. Range 1-2. Grassy swamps. Frequent; ascending from the shore swamps at Flookborough to 400 yards in Troutbeck Valley and 500 yards in Kirkstone Pass.

Astragalus hypoglottis, L.

C. Mr. W. B. Waterfall informs me that he has a specimen gathered in 1864 on Catlands or Carrock Fell, but I have no other record of its occurrence.

286. *Astragalus glyciphyllos*, L. (Wild Liquorice). Native. Germanic type. Range 1.

L. Near the Physic Well at Cartmel. First recorded by Wilson. Recent confirmation wanted.

291. *Ornithopus perpusillus*, L. (Bird's-foot). Native. British type. Range 1. Sandy ground. Not infrequent.

C. Cowrake Quarry (red sandstone) near Penrith Beacon, 300 yards.—(B.) Near Roughton, Ennerdale.—(W. Hodgson.) Middletown, Gill.—(Whitehaven Cat.) Barrow Island, Derwentwater, and in Borrowdale near Rosthwaite.—(C. Bailey.) Ravenglass, Irton, Braystones, and St. Bees Moor. First recorded by Lawson.

W. Ascending Nab Scar from Rydal Mount.—(T. J. Foggitt.) Drawn from this locality by Miss Wilson. Tenterfell, Kendal.—(Wilson.) Sandstone quarries at Common Holme Bridge.—(Lawson.)

L. East side of Coniston Lake.—(Linton.) Between Windy Ash and Higher Laith, Ulverstone.—(Miss Hodgson.)

293. *Hippocrepis comosa*, L. (Horse-shoe Vetch). Native. English type. Xerophilous. Range 1. Limestone hills. Locally plentiful.

W. Rocks by the stream that runs from Anna-well to Shap. —(Lawson.) Ledge of the scar between Scout Stile and Honey-bee Gate, Kendal. First recorded by Wilson.

L. Birk Fell, Humphrey Head, and Yewbarrow near Grange. —(W. Foggitt, C. Bailey, etc.) Copse Head near Holker, and limestone pastures in Furness.—(Aiton.)

Onobrychis sativa, L. (Sainfoin). Alien?

C. Nethertown near Egremont.—(J. Robson.)

295. *Vicia Orobus*, DC. Native. Scottish type. Range 1.

C. Hedges and pastures about Gamblesby, six miles northeast of Penrith, plentifully. Discovered by Willisel. Still there.

296. *Vicia sylvatica*, L. (Wood Vetch). Native. Scottish type. Range 1. Woods and thickets. Rare.

C. Coast cliffs at Parton and between Whitehaven and St. Bees.—(J. Woods, etc.) Seacliffs, Parton, Barrowmouth, and other places.—(Whitehaven Cat., W. Hodgson.) Clifton Woods near Workington.—(W. Dickinson.) Isell Woods near Cockermouth.—(Rev. J. Dodd.) Flimby Wood near Maryport ; also in Dentonside Wood, near the old Sebergham colliery.—(W. Hodgson.)

W. Near the bridge at Kirkby Lonsdale, where Lawson showed it to Dr. Richardson. Casterton Woods. —(Hindson.) Laverock Bridge and 'Barrowfield Wood near Kendal. First recorded by Woodward.

L. Urswick Wood and between Stoneylands and Newton-in-Cartmel.—(Aiton.)

297. *Vicia Cracca*, L. (Blue Vetch). Native. British type. Range 1-2. Hedges and meadows. Frequent; ascending to 350 yards.

298. *Vicia sativa*, L. (Common Vetch). Native. British type. Range 1. The type frequent in cultivated fields, ascending to 300 yards near Shap.

Var. *segetalis* seen as a corn-field weed about Penrith and Ulverstone, and drawn from Troutbeck by Miss Wilson.

Var. *angustifolia*, rarely truly wild, in dry grassy places. Railway cutting at Aspatria, plentiful; more sparingly about Hurrock Wood and Lake foot, Ullswater.—(W. Hodgson.)

301. *Vicia sepium*, L. (Hedge Vetch). Native. British type. Range 1-2. Thickets and rocky places. Frequent. Ascending to 560 yards.—(Watson.)

303. *Vicia hirsuta*, Koch. (Hairy Tare). Colonist. British type. Range 1. Cultivated fields. Frequent; ascending to 250 yards over Bowness. A glabrous variety in Meathop Moss.

304. *Vicia tetrasperma*, Moench. (Smooth Tare). Colonist. English type. Range 1.

C. Near Clea Hall.—(Rev. R. Wood.)

L. Seen only once, by the side of the road ascending from the Ferry Inn at Windermere to Sawrey.—(B.) Included in Aspland's list of Grange plants.

306. *Lathyrus Nissolia*, L.

C. In sandy ground at Irton.—(W. Dickinson.) Confirmation wanted.

308. *Lathyrus pratensis*, L.   Native.   British type. Range 1-2. Meadows and hedge-banks. Frequent. Ascending to 350 yards over Haweswater.—(Watson.)

309. *Lathyrus palustris*, L.

C. St. Bees.—(Whitehaven Cat.) Confirmation wanted.

310. *Lathyrus sylvestris*, L. (Everlasting Pea). Native. English type. Range 1.

C. Rocks by the Red Neese, Whitehaven. First recorded by Lawson. Rocks near Parton and between Parton and Harrington.—(Whitehaven Cat., W. Hodgson.) Erroneously referred to *latifolius* in the Botanist's Guide. Should be in some of the old Floras.

311. *Lathyrus maritimus*, Bigel. (Sea Pea). Native. Doubtful type. Maritime. Range 1.

C. Harrington rocks, between Workington and Whitehaven.—(W. Dickinson, J. Robson.) Cliff at St. Bees.—(Miss Edmunds.)

312. *Orobus tuberosus*, L. Native. British type. Range 1-2. Woods and thickets. Frequent; ascending to 350 yards in Great Langdale. 600 yards.—(Watson.)

Var. *tenuifolius*, rare in High Furness.—(Aiton.) A white-flowered variety is found in a hedgerow near the Tongue, Watermillock.—(W. Hodgson.)

ORDER ROSACEÆ.

314. *Prunus spinosa*, L. (Sloe). Native. British type. Range 1-2. Woods and hedges. Common throughout the lower zone. Ascending to 300 yards over Coniston, and 400 yards in Troutbeck Valley.

Var. *insititia* (Bullace). Frequent in hedges.

Var. *domestica* (Wild Plum). Half wild in many places, as in Bowness Woods, and near Barnbeck farm, Furness.

315. *Prunus Padus*, L. (Heckberry, Bird Cherry). Native. Scottish type. Range 1-2. Common everywhere in woods and hedges. One of the great ornaments of the Lake country in spring, and often used for arches or bouquets for wedding decorations. Ascends to 400 yards in Troutbeck Valley, and nearly as high about Watendlath.

316. *Prunus Cerasus*, L. Denizen. English type. Range 1.

W. The least wild cherry-tree, vulgarly called the Merry tree; Rossgill.—(Ray, in Gibs. Camden, p. 818.) Spital Wood, Kendal.—(T. Gough.)

L. Occasionally seen in hedges in Furness, but rare.—(Miss Hodgson.)

316. *Prunus Avium*, L. (Wild Cherry). Denizen. English type. Range 1. Everywhere common in woods and thickets through the lower zone, ascending to the summit of Yewbarrow behind Grange-over-Sands, and nearly to the top of Brant Fell over Bowness. I never saw the wild cherry so plentiful in any other part of England as at the Lakes. The wild black Martindale cherries, Mrs. King tells me, are regularly sold in Penrith market. 'The fruit,' Mr. Hodgson writes, 'has a high reputation, and within the period of my remem-

brance the three first Sundays of August were called "Cherry Sundays," and crowds of people assembled to regale themselves, and go out boating on the Lake. At Warnel Hall near Sebergham, once the property of the Dentons, stands a fine row of black cherry-trees, in comparison with which the Martindale trees are dwarfs. Fruit from these is gathered in large quantities and sent to Lowther Castle to be made into wine.'

317. *Spiræa Ulmaria*, L. (Meadow Sweet). Native. British type. Range 1-2. Woods and damp places. Common. Ascending to 500 yards in Kirkstone Pass. 560 yards.—(Watson.)

318. *Spiræa Filipendula*, L. (Dropwort). Native. Xerophilous. English type. Range 1. Dry banks on the limestone. Rare.

C. St. Bees.—(Whitehaven Cat.) Meadows at Gosforth.—(J. Robson.) Middletown Hamlet near St. Bees.—(W. Hodgson.)

W. Cunswick Scar and Barrowfield Wood near Kendal. First recorded by Lawson.

L. Humphrey Head, on the top, with *Helianthemum canum*. —(Dr. Windsor.) Abundant at Arnside.—(J. C. Melvill.)

Spiræa salicifolia, L. Alien. An occasional straggler from cultivation. The best-known station is by the side of the road between Colthouse and Hawkshead, a little above the head of Esthwaite Water, where it was first noticed by the celebrated John Dalton the chemist, and communicated to Withering. It grows also on the shore of Windermere near the Ferry Inn, and of Coniston Lake near Waterhead. Mr. Hodgson pointed out to me several bushes of *S. hypericifolia* on a bank over Ullswater near the Pooley Bridge landing-stage.

321. *Geum urbanum*, L. (Wood Avens). Native. British type. Range 1-2. Woods and hedge-banks. Common in the

lower zone, ascending to 560 yards. *G. intermedium*, a variable hybrid between the two species, is plentiful in woods about Ullswater, Lowther, Keswick, Arnside, etc.

322. *Geum rivale*, L. (Water Avens). Native. British type. Range 1-3. Damp and shaded woods, and rocky places. Common; ascending to 850 yards on Scawfell Pikes, where it overlaps *Salix herbacea*.—(Watson.) A proliferous monstrosity has been gathered by Lawson at Great Strickland, and Mr. C. Bailey in Borrowdale. I have often seen the variety described by Lawson near the river Ellen from Aspatria downwards. It is a handsome plant.—(W. Hodgson.)

323. *Dryas octopetala*, L. Native. Highland type. Range 3.

W. On Helvellyn above Keppel Cove Tarn.—(J. Backhouse.)

325. *Potentilla fruticosa*, L. (Shrubby Cinquefoil). Native. Intermediate type. Range 1-3.

C. In a ravine of the Wastdale Screes, called the Devil's Sled-gate. First recorded by Bicheno and Woods.

W. On Helvellyn above Keppel Cove Tarn.—(J. Backhouse.)

L. Has been found near Ulpha.—(Aiton.)

327. *Potentilla anserina*, L. (Silver Weed). Native. British type. Range 1. Roadsides and waste ground. Common; ascending from the coast-level at Flookborough to 250 yards in Troutbeck Valley, and 300 yards at Shap and over Penrith.

328. *Potentilla argentea*, L. Native. English type. Range 1.

L. On the shore of the Leven estuary near Conishead bank.—(Aiton.)

329. *Potentilla verna*, L. Native. British type. Range 1. Limestone rocks. Very rare.

W. On Whitbarrow.—(F. Clowes.)

L. Near Grange-over-Sands.—(Rev. H. Higgins.) Miss Beever gathered it in Silverdale, just beyond our limits.

330. *Potentilla alpestris*, Hall. fil. Native. Highland type. Range 2.

W. Rocks on the ascent of Grange Fell from the Vale of Newlands, 350 yards.—(Watson.)

331. *Potentilla reptans*, L. (Creeping Cinquefoil). Native. English type. Range 1. Roadsides and hedge-banks. Frequent in the lower zone, ascending to 300 yards near Shap.

332. *Potentilla Tormentilla*, Schreb. (Common Cinquefoil). Native. British type. Range 1-4. Heathy places at all elevations. Common; ascending to 850 yards on Grisedale Pike, and 1000 yards on Skiddaw.

Var. *procumbens*, Sibth. At Lowwood, Stock Ghyll, Coniston, Arnside, and many other places.

333. *Potentilla Fragariastrum*, Ehrh. (Barren Strawberry). Native. British type. Range 1-2. Hedge-banks and woods. Frequent; ascending to the limestone pavement of Whitbarrow, to 400 yards in Great Langdale, 500 yards between Borrowdale and Thirlmere; 600 yards (Watson).

334. *Comarum palustre*, L. (Marsh Cinquefoil), Native. British type. Range 1-2. Common in peat-bogs; ascending from coast-level at Plumpton near Ulverstone to 500 yards on the Stake Pass.

335. *Fragaria vesca*, L. (Wood Strawberry). Native. British type. Range 1-2. Common in woods, and on hedge-

banks, ascending to the limestone pavement of Hutton-Roof Crags, and to 400 yards in Great Langdale.

Fragaria elatior, Ehrh. (Hautboy Strawberry). Alien. Formerly much cultivated, but now replaced by the Chilian species, *F. chilensis.*

C. Woodhall near Keswick.—(W. Dickinson.) Roadside at Barrow, and near a farm-house on the east slope of Latrigg. —(B.) In Borrowdale near Grange.—(C. Bailey.)

W. Side of the main road near Bowness.—(B.) Railway bank between Grayrigg and Oxenholme.—(Britten and Holland.)

L. By a brook in the Vale of Newland near Ulverstone.—(Miss Parker.)

337. *Rubus Chamæmorus*, L. (Noutberry, Cloudberry). Native. Highland type. Range 3. Common on the millstone grit summits of the Pennine chain, where its lower limit marks explicitly the line of boundary between Watson's Agrarian and Arctic regions, but very rare, and local in the Lake district.

C. In Styx Moss at the head of the Glencoin Valley.—(W. Dickinson.) High Pike.—(Rev. R. Wood.)

W. On the Shap Fells between Banisdale Head and Wet Sleddale.—(Lawson.) High Street, Goat Scar, and Long Sleddale.—(F. Clowes.) On the top of a high mountain called Goatcow over Long Sleddale.—(Wilson.) The Kendal stations given by Martyn belong to *R. saxatilis.*

338. *Rubus saxatilis*, L. Native. Scottish type. Range 1-3. Woods and cliffs at all elevations. Not infrequent.

C. Cockshot Wood and Walla Crag near Keswick.—(W. Robertson, J. B. Davies.) Erne Crag and Great Crag between Borrowdale and Thirlmere.—(Watson.) Ravine of Aira Beck above Dockray.—(W. Hodgson.) Cliffs of Great End above Sprinkling Tarn, 800 yards.—(Watson.) Banks of Aira Beck. —(Whitehaven Cat.)

W. Cliffs of the east face of Helvellyn.—(Balfour.) In a few places round Windermere.—(F. Clowes.) Cunswick Wood and other places near Kendal.—(T. Gough.) Abundant with *Convallaria majalis* in Middlebarrow Wood, Arnside.—(B.) Barrowfield Wood and Caldkale Scroggs near Kendal.—(Wilson.) Lowther Park.—(Rev. A. Ley.) Underbarrow Scar, Kendal, and by the stream descending from Scandale to Patterdale.—(J. Ball.)

L. Humphrey Head.—(Dr. Windsor.) Rowdsey Wood near Haverthwaite.—(Miss Hodgson.)

339. *Rubus Idæus*, L. (Raspberry). Native. British type. Range 1-2. Woods and thickets. Common ; ascending to 400 yards in Great Langdale ; 500 yards by the waterfall on the south-west slope of Saddleback (Watson); and as high as the hills round Ullswater.

Var. *Leesii*, Bab., was found by Mr. Edwin Lees, after whom it is named, on the banks of the stream that runs into Windermere between Bowness and Troutbeck. I sought for it there without success in 1883.

340—3. *Rubus suberectus*, And. (Bramble, as are named the other sub-species of *fruticosus*). Native. British type. Range 1. A widely-spread Lake species. Frequent about Keswick and Ambleside, ascending into Great Langdale and the Watendlath valleys, and seen also at Coniston, Grange-over-Sands, Meathop Moss, in the Duddon valley at Seathwaite, the Troutbeck valley, the Vale of St. John, Haweswater, etc. The three varieties, *suberectus*, *fissus*, and *plicatus*, all occur in the district.

340—6. *Rubus affinis*, Bab. A universally distributed

Lake sub-species. Keswick, Watendlath, Borrowdale, Ambleside, Haweswater, Grasmere, Wastdale Head, Coniston (where it ascends to 300 yards), Valley of St. John, Witherslack, Grange-over-Sands, Newby Bridge, Ulverstone, Great Strickland, etc. I have gathered a plant that matches the well-known *R. laciniatus* of gardens in hedges near the Post Office at Grasmere, and a form with less decidedly laciniated leaves and ascending sepals between Ulverstone and Swarthmore Hall. I have dried specimens of all the Lake forms of fruticose *Rubi* and placed them in the Kew Herbarium.

340—7. *Rubus Lindleianus*, Lees. A universally distributed Lake sub-species. Keswick, ascending Borrowdale to Seatollar, Ambleside, Bowness, and high up the Troutbeck valley, Sawrey, and woods about the Ferry Inn, Coniston village, Witherslack, Newby Bridge, Arnside, Watermillock, Grange-over-Sands, Humphrey Head, Cartmel, Ulverstone, etc.

340—8. *Rubus rhamnifolius*, W. and N. Widely spread, but not so common at the Lakes as *affinis*, *Lindleianus*, *umbrosus*, and *pallidus*, which are the four most predominant bramble types.

Var. *cordifolius* abundant about Keswick, especially in the lane leading towards Skiddaw from the railway station; also in the Witherslack valley, and about Coniston, and in Tilberthwaite Ghyll. The smaller-leaved finely-serrated typical *rhamnifolius* in the lane near the Druidic Circle at Keswick, Lodore Woods, and in Borrowdale, at Stonethwaite, and between Grange and Castle Crag. I have seen the sub-species also at Sawrey, Holme Mill, high up Troutbeck, at Clibburn, Lowther, and in the Vale of St. John. There is a form with leaves densely hairy beneath, very near the Llanberis *incurvatus*, below Watermillock, where it was shown me by Mr. W. Hodgson, and in several other places round Ullswater, and I have seen it also at the foot of Haweswater.

340—12. *Rubus discolor*, W. and N. Not seen in the heart of the Lake district about Keswick, Ambleside, Bowness, and Coniston, but plentiful enough on the outskirts, in Furness and about Arnside, Kirkby Lonsdale, Milnthorpe, and Penrith. I have seen a form with laciniated leaves at Ulverstone on the way to Swarthmore, and good var. *pubescens*, Bab., which connects this with *leucostachys*, near Arnside Tower. *R. thyrsoideus* I have never seen within our boundaries.

340—14. *Rubus leucostachys*, Sm. The commonest form in all the woods and hedges about Arnside, and also common about Milnthorpe, Newton, Newby Bridge, Witherslack, and Grange-over-Sands. Seen also at Coniston, Bowness, Grasmere, Colwith Force, Cartmel, Ulverstone, Sawrey, in the Vale of Lorton near Brackenthwaite, and in Borrowdale between Seatollar and Seathwaite.

340—17. *Rubus Salteri*, Bab.

Var. *R. calvatus*, Blox.

W. Hedges at Holme and Burton in Lonsdale, and near Milnthorpe station. Hillside over Bowness and up the Troutbeck valley. A little out of Ambleside on the Kirkstone road. Witherslack Valley between Witherslack Hall and Townend. I find much difficulty in drawing the line between this and some of the forms of *rhamnifolius*.

340—18. *Rubus carpinifolius*, Bab. This, which is a plant I do not understand, is given by Professor Babington as occurring near Keswick and in Stock Ghyll.

340—19. *Rubus villicaulis*, W. and N. I have notes of two plants that probably should range here, one a large coarse form, growing in Borrowdale in hedges north of Rosthwaite, and the other a form with more finely cut leaves between

Ritson's inn and the head of Wastwater. *R. mucronatus*, which is tolerably frequent in North Yorkshire, I have not seen anywhere at the Lakes.

340—20. *Rubus macrophyllus*, W. and N. I have not seen typical *macrophyllus* at the Lakes, but Professor Babington mentions it as found at Ambleside and Bowness. *R. umbrosus*, Arrh. (*R. carpinifolius*, Blox.), is one of the most universal Lake *Rubi*, ascending to 250 yards in the Watendlath valley and 300 yards in Great Langdale. A form between *umbrosus* and *pyramidalis* is found in the woods about Barrow and ascending the hill from Windermere ferry to Sawrey.

340—22. *Rubus Sprengelii*, W. and N. Very rare at the Lakes. Given by Professor Babington as an Ambleside plant. I have only once seen it, by the side of the road by Windermere, south of Storrs Hall.

340—28. *Rubus rudis*, W. and N. Very rare at the Lakes. Gathered by Dr. Cookson near Rydal. I have seen it once only, at Clappersgate near Ambleside. *R. Bloxami*, *scaber*, and *Hystrix* are not known at the Lakes.

340—29. *Rubus Radula*, W. and N. Rare in the interior of the Lakes.

C. Hedges near the Druidic Circle, Keswick. Lodore: two forms near the stream below the wood, and another on the hillside above High Lodore.

W. In several places between Shap and Clifton, and about Clibburn and Great Strickland. Hedges between Holme mill and the railway station.

L. Roadside at Newton and Newby Bridge. Brick-kiln and plantations at Ulverstone.—(Miss Hodgson.)

340—30. *Rubus Koehleri*, W. and N.

Var. *pallidus* is one of the commonest Lake *Rubi*, occurring especially in woods. It ascends to the top of Whitbarrow and Hutton-Roof Crags, and to 300 yards over Coniston. A variety with adpressed sepals, first noted by Turner, occurs in Rydal Woods and in Langstrathdale above Stonethwaite. Fine var. *infestus* grows in the hedges of the Keswick and Penrith road near the Druidic Circle. Typical *Koehleri* is reported by Professor Babington from Stock Ghyll and Rydal, and I have seen it in hedges near Arnside Station and high up the Troutbeck valley.

340—32. *Rubus dumetorum*, W. and N. Very rare at the Lakes.

W. Roadside near Storrs Hall, Windermere. Hedge between Holme Mill and the railway station.

L. Hedges of the upper road between Grange and Lindale. Shore near Kents Bank railway station.

340—35. *Rubus rosaceus*, W. and N.

C. Abundant in the lane leading up Skiddaw from Keswick station. Lane at Wastdale Head, west of Ritson's inn.

W. Rydal Falls, *teste* Babington. Roadside near Brathay Bridge. Bowness Woods. Hill west of Witherslack Hall.

L. Woods round the Ferry Inn, Windermere.

340—38. *Rubus Bellardi*, W. and N.

L. Edge of Longhause Ghyll, Walna Scar, about a mile from Seathwaite Church.—(Miss Hodgson.) *R. pyramidalis* and *Guntheri* have never been found at the Lakes.

340—40. *Rubus corylifolius*, Sm. Rare in the heart of the Lake country about Ambleside and Keswick, but common on the outskirts about Cockermouth, Ulverstone, Grange, Shap, and Penrith. Ascends to 300 yards on Shap Common, and nearly as high at Rossgill.

Vars. *degener* and *Balfourianus* both occur.

340—43. *Rubus cæsius*, L. (Dew Berry). Native. English type. Range 1. Like *corylifolius*, rare in the heart of the Lake country, but common in many places on the outskirts, as about Cockermouth, Grange, Arnside, Penrith, Lowther, and the Winster valley. It ascends to the limestone pavement of Hutton-Roof. About Gilcrux, Plumbland, Torpenhow, and Ireby, on limestone, this is the master bramble.— (W. Hodgson.)

Var. *pseudo-idæus*, near St. Paul's Church in the Winster valley.

341. *Rosa spinosissima*, L. (Scotch Rose, Burnet Rose). Native. British type. Range 1. Frequent along the coast sand-hills, as, for instance, at Seascale, where it covers a wide area. Inland, in many places round Windermere, Derwentwater, Haweswater, and Ullswater. Abundant on Whitbarrow, Arnside Knot, and the other limestone hills, ascending to 300 yards near Shap.

342. *Rosa hibernica*, Smith (Irish Rose). Native. Intermediate type. Range 1.

C. In Lorton Vale in hedges and thickets for about two miles on both sides of Lorton village. Discovered by Borrer in 1845.

Rosa lucida, Ehrh. Alien. Two bushes in the hedge near the junction of the Greta and Derwent at Howray, Keswick. —(Borrer.) Given in several of the guide-books as *R. cinnamomea*.

343. *Rosa Sabini*, Woods. Native. British type. Range 1.

C. By the side of the road ascending Whinlatter from Braithwaite, between three and four miles from Keswick. Discovered by J. Woods in 1800. His *R. gracilis* was founded on this and the Pooley Bridge plant. West side of Derwentwater opposite St. Herbert's island.—(W. Dickinson.) Roadside near both the entrances to Dalemain Park.—(B.) Hedges between Lamplugh Cross and Ennerdale.—(Winch.)

W. A few plants near Pooley Bridge, and plentiful in the direction of Lowther and Howtown. First noted by Woods in 1808. Near Haweswater.—(J. Woods.)

L. Abundant in several places about Cartmel.—(J. Sidebotham.)

344. *Rosa mollis*, Smith. Native. British type. Range 1-2. Woods and thickets, common in the lower zone, ascending to 420 yards.—(Watson.) Very fine about Keswick, Bowness, Penrith, Pooley Bridge, Shap, and between Haweswater and Rossgill, and very ornamental when the bright red globes of fruit are ripe in September.

Var. *cærulea*, Woods. In hedges at Lowther, Howtown, etc.

345. *Rosa tomentosa*, Smith. Native. British type. Range 1. Hedges and thickets, not so common at the Lakes as *canina* and *mollis*; ascending to 300 yards in Langdale and over Coniston.

Var. *scabriuscula* is not infrequent. Fine var. *sylvestris* grows in the lane leading up Skiddaw from the Keswick railway station. A form near var. *farinosa* was found by the Rev. A. Ley at Portinscale.

Rosa rubiginosa, L. (Sweet Briar). Alien.

C. Ennerdale and Kinniside, frequent. — (Whitehaven Cat.) In two places a mile apart, in hedges of the main road at Shafton near Cockermouth.—(B.) Hedgerow near Watermillock Church, introduced.—(W. Hodgson.) Brayton.-- (W. B. Waterfall.)

L. Occasionally in hedges and thickets in Furness.—

(Aiton.) *R. micrantha* and *inodora* are not known within our limits.

351. *Rosa canina*, L. (Dog Rose). Native. British type. Range 1-2. Everywhere common in woods and hedges in the lower zone, ascending to 400 yards in Great Langdale. The characteristic feature of the Lakeland Dog Roses is the plenty and luxuriance of the varieties of the subcristate series, especially *Reuteri*, *subcristata*, *coriifolia*, and *Watsoni*. The three forms described by Woods, *nuda* and the two varieties of *bractescens*, were founded on Lake examples. Besides these, I have seen fine *marginata* in the lane between Newby Head and Common Holme Bridge. Of the forms with deciduous sepals, *lutetiana*, *sphærica*, *dumalis*, and *urbica* are common, and Miss Hodgson and I both found *frondosa* in Furness, and I have seen *biserrata* in hedges near Lowther. I have not seen *tomentella*, nor good *dumetorum*, nor *arvatica*, nor characteristic plants of any of the Hispidæ.

353. *Rosa arvensis*, Huds. (York Rose). Native. English type. Range 1. Not seen about Keswick, Penrith, Shap, Ambleside, or round Ullswater, but it occurs at the foot of Windermere near Newby Bridge, and at Coniston, and is abundant in the Witherslack valley and about Grange-over-Sands, Arnside, and Whitehaven.

354. *Agrimonia Eupatoria*, L. (Agrimony). Native. British type. Range 1. Roadsides and grassy banks, frequent, ascending to 250 yards at Shap and in Troutbeck Valley.

354*. *Agrimonia odorata*, Ait. Native. British type. Range 1.

C. Dent Hill.—(Rev. F. Addison.) Vale of Lorton.—(W. Robinson.) Troutbeck, High Lodore, and a little out of Keswick on the Ambleside road.—(B.)

W. In a wood near the railway station at Arnside.—(W. Foggitt.)

L. Shore of Windermere near the Ferry Inn.—(B.)

354*. *Sanguisorba officinalis*, L. (Great Burnet). Native. Intermediate type. Range 1. Damp meadows, sometimes even in corn-fields. More frequent at the Lakes than I have anywhere else seen it, ascending to 450 yards.—(Watson.)

355. *Poterium Sanguisorba*, L. (Lesser Burnet). Native. English type. Xerophilous. Range 1. Rocks and dry banks, almost confined to the limestone.

C. Snebra near Whitehaven and Eskdale.—(Whitehaven Cat.) Railway banks a mile from Penrith on the Keswick line.—(B.)

W. Banks of the Leith at Clibburn.—(B.) Shap Common and Lowther Woods, ascending to 300 yards. Moordivock Quarries over Askham.—(W. Hodgson.) Frequent on the limestone hills south of Kendal, ascending to the summit of Whitbarrow.—(B.) Abundant about Arnside.—(B.) Frequent at Kirkby Lonsdale.—(Hindson.) Roadside south of Bowness, on slate.—(B.)

L. Islands of Windermere.—(W. Foggitt.) Humphrey Head and other hills in Furness.—(Aiton, B.) Plumpton Quarry near Ulverstone, and roadsides above Grange.—(Miss Hodgson.)

356. *Alchemilla vulgaris*, L. (Lady's Mantle). Native. British type. Range 1-3. Meadows and grassy banks. Frequent; ascending to the limestone pavement of Hutton-Roof, 850 yards on the Striding-edge Crags, and overlapping *Carex rigida* on Scawfell Pike. Hudson founds a variety from the Westmoreland mountains on Plukenet's *Alchemilla minor*, figured in Phytographia, tab. 240, fig. 1.

357. *Alchemilla alpina*, L. (Alpine Lady's Mantle). Native. Highland type. Range 1-3. The abundance of this plant, which is absent from Northumberland and Durham, and only found in one place in Yorkshire, where the Silurian rocks just pass within the bounds of the county, is one of the most characteristic notes of the Lake flora. It is quite confined to the slate hills, being plentiful on Scawfell, Scawfell Pikes, Great End, Lingmell, Great Gable (where it ascends to 950 yards), Red Pike, Pillar, over Blacksail Pass down to the foot of Wastwater, and in Ennerdale down to the bridge over the Liza. It is abundant on Honister Crag and round Keswick, reaches down to the Vale of Newlands and the foot of Castle Crag. In Westmoreland it is plentiful on Helvellyn, Fairfield, High Street, and the hills round Ullswater and Haweswater, reaching down to the stream of the Vale of St. John at the foot of Great Dod. It is absent from the limestone hills south of Kendal, and is omitted from Miss Hodgson's flora of Furness. In Journ. Bot. 1872, p. 308, Rev. R. Wood refers a Cumbrian plant, gathered by Mr. W. Dickinson, to *A. conjuncta*, Bab., which I believe to be merely a variety of *alpina*.

358. *Alchemilla arvensis*, L. (Parsley Piert). Native. British type. Range 1-2. Cultivated fields and wall-tops. Frequent. Ascending to 370 yards.—(Watson.)

Mespilus germanica, L. (Medlar). Alien.

L. Two trees in a high old hedge between Walney Church and North Scale, east side of Walney Island.—(Miss Hodgson.)

360. *Cratægus Oxyacantha*, L. (Hawthorn). Native. British type. Range 1-2. Woods and hedges. Common everywhere in the lower zone, ascending to the limestone pavement of Hutton-Roof and Farleton Knot to 400 yards; in

Great Langdale and Troutbeck Valley to 450 yards. There is a curious wood of old hawthorns, with trunks overgrown with *Usnea*, by the side of the Roman road at the foot of Hill Bell. All the Lake plant is *C. monogyna*, Jacq.

Var. *laciniata* I have seen in hedges at Arnside and several other places.

Pyrus communis, L. (Wild Pear-tree). Alien.

C. Crag Farm, Watermillock, and a fine tree in a hedgerow at Gatesgill near Carlisle.—(W. Hodgson.) Curthwaite near Wigton.—(Rev. R. Wood).

W. Roadside between Townend and Witherslack.—(B.)

L. Hedges near Ulverstone.—(Aiton.) A fine tree in a hedge facing the Morecambe shore at Bardsea.—(Miss Hodgson.)

363. *Pyrus Malus*, L. (Crab-tree). Native. English type. Range 1. Hedges and thickets. Frequent throughout the lower zone, ascending to 250 yards in Naddle Forest over Haweswater, and 300 yards on Shap Common. I have notes of var. *tomentosa* from hedges east of Greystoke, and between Newby and Common Holme Bridge, and Miss Hodgson cites several stations in Furness.

Pyrus torminalis, Ehrh. Alien.
W. Levens Park, near the bridge.—(Lawson.)
L. Plumpton Woods, Ulverstone.—(Aiton.)

365. *Pyrus Aria*, Sm. (White Beam Tree). Native. English type. Xerophilous. Range 1. Confined to the limestone hills and cliffs, where it is widely spread. Scout Scar and Cunswick Scar near Kendal, on Whitbarrow, ascending to the limestone pavement of the summit; and at Arnside, Meathop, and Humphrey Head, down to the shore cliffs. The type and var. *rupicola* both occur. *P. semipinnata*,

Roth., was distributed by the Rev. A. Ley through the Exchange Club from planted woods at Wastdale Head.

366. *Pyrus Aucuparia*, Gaertn. (Rowan, Mountain Ash). Native. British type. Range 1-3. One of the commonest and most ornamental trees of the Lake district, both amongst the slate and limestone hills, ascending higher than any other tree except the juniper. I have seen it at 900 yards on the Striding-edge Crag.

ORDER ONAGRACEÆ.

367. *Epilobium angustifolium*, L. (Rose-bay Willow Herb.) Native. British type. Range 1-2. Woods and cliffs. Not infrequent.

C. Abundant on the coast cliffs near St. Bees Head.—(Rev. F. Addison.) In Borrowdale on the hill-side east of Rosthwaite.—(W. Foggitt.) Ascent of Grange Fell from the Vale of Newlands.—(Watson.) Church Moss, Beckermet and Drigg.—(Whitehaven Cat.) Great Mell Fell.—(W. Hodgson.) Hobcarten Crag, 1500 feet.—(W. B. Waterfall.)

W. Highbarrow Bridge near Shap.—(Linton.) Plantation near Hackthorp.—(B.) Ravine of Swarth Fell. Rocky bluff above Otterstone by Ullswater.—(W. Hodgson.) Head of Long Sleddale.—(T. Gough.) High rocks above Sweden Bridge, Ambleside, on the way to Patterdale.—(Miss Beever.) Rocks in Mardale, 450 yards.—(Watson.)

L. Hawkshead.—(Martyn.) Conishead Priory, and riverside between Cark and Cartmel.—(Aiton.) By the beck at Newlands, perhaps introduced.—(Miss Hodgson.)

368. *Epilobium hirsutum*, L. (Great Willow Herb, Apple Pie). Native. English type. Range 1. Bogs and streamsides. Frequent in the lower zone, ascending to Watermillock, 300 yards.—(W. Hodgson.) Plentiful about Aspatria,

Clibburn, Great Strickland, Penrith, Kendal, Carlisle, Cockermouth, etc.

369. *Epilobium parviflorum*, Schreb. Native. British type. Range 1. Stream-sides and swamps. Frequent through the lower zone, ascending to 300 yards near Shap, and the Walna Scar slate quarries.

Var. *rivulare* in ditches near Newby, and shown to me by Mr. W. Hodgson in a bog east of Watermillock over Ullswater.

370. *Epilobium montanum*, L. (Common Willow Herb). Native. British type. Range 1-2. Woods and shaded rocks. Frequent; ascending to the limestone pavement of Hutton-Roof and Farleton Knot; 560 yards.—(Watson.) I have no knowledge of *E. roseum* within our limits. *E. lanceolatum* of the Whitehaven list is no doubt a misnomer.

372. *Epilobium palustre*, L. (Bog Willow Herb). Native. British type. Range 1-2. Swamps and ditches, ascending from shore level in peat-mosses at Ulverstone to 640 yards. —(Watson.)

373. *Epilobium obscurum*, Schreb. Native. British type. Range 1. Ditches and swamps. Frequent in the lower zone; ascending to 250 yards over Lodore and in Troutbeck Valley, to 300 yards in Mardale.—(Watson.) Typical *tetragonum* not known.

374*. *Epilobium alsinifolium*, Vill. (Mountain Willow Herb). Native. Highland type. Range 1-3.

C. Helvellyn and Great Dod, from the highest springs down to bottom of the Vale of St. John. Sykes of Styhead Pass down the slope of Great Gable, and in the Cocker valley at Whinlatter.—(B.)

W. Springs of west side of Helvellyn down into Tongue Gill.—(B.) Shap, Mardale, and abundant high up in Kentmere and Long Sleddale, gathered long ago by Curtis (who published it as *alpinum*) and Woods. High Street, down into Troutbeck and Kirkstone Pass.—(B.) Not known in Lake Lancashire. The true *E. alpinum*, lately discovered by Rev. F. Addison on Cross Fell, is not known at the Lakes.

Isnardia palustris, L.

C. In the Liza in Ennerdale.—(J. Robson.) No doubt a misnomer. *Peplis* probably intended, which grows in the same stream.

Oenothera biennis, L. (Evening Primrose). Alien.

C. Railway cutting at Dalston.—(Rev. R. Wood.)

377. *Circæa lutetiana*, L. (Enchanter's Nightshade). Native. British type. Range 1. Woods and thickets. Frequent in the lower zone, ascending to the limestone pavement of Farleton Knot, to Haweswater, Seathwaite in Borrowdale, and 300 yards in Mardale.

378. *Circæa alpina*, L. Native. Scottish type. Range 1-2. Woods and thickets. Frequent, both the type and var. *intermedia*, Ehrh.

C. Castlehead Wood, Ashness Gill, and other places round Derwentwater; first recorded by D. Turner. West side of Thirlmere.—(Rev. A. Ley.) In the Cocker valley at Lorton. —(Harriman.) Rocks on the slope of Haystacks over the Seatollar and Buttermere road, up to 500 yards.—(B.) Stybarrow Crag and Ullswater shore at Waterfoot.—(B.) In Ennerdale, between the lake and Ennerdale Bridge.—(W. Foggitt.) Buttermere, in hedges near the village.—(Rev. A. Ley.)

W. Mardale and Shap, both type and var. *intermedia.*—

(Watson.) By Stickle Tarn, at the foot of Langdale Pikes.—
(J. C. Melvill.) Dungeon Ghyll, in the ravine between the
two waterfalls.—(B.) Shaded walls at Bowness.—(B.) Near
Milnthorpe at Storth and Dallam Tower.—(E. Robson, T.
Gough.) Foot of Wansfell and back of Silver Howe.—
(F. C. Roper.) Drawn from Fox Ghyll by Miss Wilson. In
several places on the stony shore of Windermere (*intermedia*).
—(J. Ball.)

L. Shore of Coniston Water and walls by the side of the
main road in Coniston village.—(Miss Beever, B.) Several
places about the road between Hawkshead and Ulverstone.—
(Mr. Crowe.) Not seen in Furness, or about Arnside, or
anywhere amongst the limestone hills.

ORDER HALORAGIACEÆ.

379. *Hippuris vulgaris*, L. (Mare's Tail, Paddock Pipe).
Native. British type. Range 1. Ponds and ditches. Not
infrequent.

C. In Dub Beck near Cleator.—(J. Robson.) Mossy
watercourse near Dubmill, Allonby.—(W. Hodgson.)

W. Pond at Askham near Lowther.—(T. J. Foggitt.)
Abundant in the two reservoirs at Holme Mill, where it is
recorded by Martyn.—(B.) Brigstear Moss, Kendal; first
recorded by Curtis. Ditches by the railway side below
Middlebarrow Wood, Arnside.—(B.)

L. In the mill-pond at Bardsea.—(Miss Hodgson.) Ditches
near Cartmel.—(C. J. Ashfield.)

380. *Myriophyllum verticillatum*, L. Native. English
type. Range 1.

C. Head of Derwentwater, at High Lodore, in ditches with
Sparganium minimum.—(B.) Naddle Beck.—(W. Dickinson.)

W. Ullswater, in a ditch near Howtown.—(W. Hodgson.)

Abundant in Windermere.—(F. Clowes.) Ditches in Brigstear Moss near Kendal.—(Wilson.)

L. Coniston Lake.—(Miss Beever.)

381. *Myriophyllum spicatum*, L. (Millfoil). Native. British type. Range 1-2. Common in ponds and lakes; ascending from Windermere and Urswick Tarn to the upper tarn at Watendlath.—(Watson.)

382. *Myriophyllum alterniflorum*, DC. Native. British type. Range 1-2. The commonest species of the tarns and peaty pools of the hill country. Mr. Borrer gathered it in Angle Tarn, Place Fell, at 500 yards, and it is abundant at the same elevation in the peaty streams in the hollow south of Hayes Water, at the foot of the precipices of High Street.

383. *Callitriche verna*, Kutz. (Starwort). Native. British type. Range 1-2. Ponds and streams. Common; ascending to Red Tarn, Helvellyn, 800 yards, and to Hayes Tarn and Low Water, Coniston Old Man.

384. *Callitriche platycarpa*, Kutz. Native. British type. Range 1-2. Ponds and swamps. Common; ascending to the top of Kirkstone Pass, 500 yards.

385. *Callitriche hamulata*, Kutz. Native. British type. Range 1-2. Ponds and streams. Frequent; ascending to Hayes Tarn, 500 yards.

C. autumnalis is recorded from Ennerdale Lake and ditches at the foot of Derwentwater, but I have never seen the true plant within our bounds.

ORDER LYTHRACEÆ.

Lythrum hyssopifolia, L.

C. Langthatch and roadside in Wastdale.—(J. Robson.) Reported in Black's Guide from the south end of Derwent water. Confirmation wanted. I suspect a mistake of name in both cases.

390. *Lythrum Salicaria*, L. (Purple Loose-strife). Native. English type. Range 1. Swamps and lakesides. Frequent; ascending to 250 yards over Coniston, and as high over Ullswater at Black Dyke, Baldhow.

391. *Peplis Portula*, L. Native. British type. Range 1. Ponds and ditches.

C. Lady Moss, Nethertown.—(Whitehaven Cat.) Harras Moor, Kinniside Long Moor, and Calder Gills.—(W. Dickinson.)

W. Ditches near Shap, 250 yards.—(Watson.) Pond at the Windermere rifle-practice ground.—(B.) End of Rydal Lake towards Grasmere.—(J. C. Melvill.) Brigstear Moss, Kendal.—(T. Gough.)

ORDER CUCURBITACEÆ.

393. *Bryonia dioica*, L. (Bryony). Native. English type. Range 1.

W. Frequent in hedges and woods at Kirkby Lonsdale.—(Hindson.) Not seen anywhere about Keswick, Ambleside, Coniston, or Penrith.

ORDER PORTULACEÆ.

394. *Montia fontana*, L. (Water Blinks). Native. British type. Range 1-3. Common in the hill springs and streamlets; ascending to 900 yards on Scawfell Pike, 800 yards on Helvellyn, 700 yards on Great Gable.

ORDER SCLERANTHACEÆ.

399. *Scleranthus annuus*, L. (Knawel). Native. British type. Range 1. Sandy ground and wall-tops. Not uncommon.

C. St. Bees, Eskdale, and Knockmurton.—(W. Dickinson.) Derwentside near Workington.—(Linton.) Nethertown near Egremont.—(J. Robson.) West Newton.—(Whitehaven Cat.) Sides of the road between Keswick and Legberthwaite.—(T. J. Foggitt.) Sandstone round Penrith Beacon, 300 yards.—(B.)

W. Between Winster and Gillhead.—(F. C. Roper.)

L. Newby Bridge.—(W. Foggitt.) Abundant on walls near the station at Woodland near Broughton-in-Furness.—(Miss Beever.) Rocky fields and wall-tops; frequent in Furness.—(Miss Hodgson.)

ORDER GROSSULARIACEÆ.

403. *Ribes nigrum*, L. (Black Currant). Denizen. Intermediate type. Range 1.

C. Wood below Penrith Beacon, and abundant on the banks of the Leith at Clibburn.—(B.) Edenside near Cotehouse.—(W. Duckworth.) By the river Eamont at Dalemain.—(T. J. Foggitt, W. Hodgson.) Mentioned as a plant of the county by Ray.

W. East side of Loughrigg Fell, Rydal.—(J. Ball.) Banks of the Mint at Kendal.—(T. Gough.) Given as a plant of the county by Hudson.

L. Roadside below Gunner's How at the foot of Windermere.—(B.) Angerton Moss and between Bowstead Gates and the Blacking Mill.—(Miss Hodgson.)

404. *Ribes rubrum*, L. (Red Currant). Denizen. Intermediate type. Range 1.

C. By Powbeck and the rivers Ehen and Calder.—(Whitehaven Cat.) Hedge near Buttermere village.—(B.) Banks of the Eamont near Dalemain.—(W. Hodgson.) Loweswater.—(W. B. Waterfall.) Hedges near Whinside Hill, Ullock Moss, and other places near Keswick.—(Watson.) Banks of the river at Threlkeld, plentiful.—(B.) Banks of the river Ellen at Baggrow near Aspatria.—(W. Hodgson.)

W. Plentiful in some of the woods round Windermere.—(F. Clowes.) Near a farm-house on the hill west of Witherslack Hall.—(B.) Frequent on the banks of the Lune at Kirkby Lonsdale.—(Hindson.) Behind Swan Inn, Grasmere. —(F. C. Roper.)

L. Newby Haw, Haverthwaite, Hearings Wood, Ulverstone, and Bank House Ghyll, Kirkby.—(Miss Hodgson.)

404*. *Ribes petræum*, Sm. Native. Scottish type. Range 1.

C. Hedges near the Vicarage, Keswick, and by the road to Wigton beyond Liswick.—(Winch.) Hedges near Ullock Moss, and lane on the right-hand side of the road from Keswick to Crosthwaite Church.—(Watson.) Roadside between Brough and Kendal, a few miles from the latter.—(Rev. J. Harriman.)

405. *Ribes alpinum*, L. Denizen. Intermediate type. Range 1.

C. Corney Fell near Ulpha.—(J. Robson.)

W. Banks of the Mint, Kendal.—(T. Gough.) Langdale. —(J. Sidebotham.)

L. Shore of Windermere near the Ferry Inn.—(T. J. Foggitt, B.) Coniston.—(Linton.)

406. *Ribes Grossularia*, L. (Gooseberry). Denizen. Inter-

mediate type. Range 1. Frequent in hedges and by roadsides and in thickets through the lower zone, ascending from Humphrey Head to the limestone pavement of Hutton-Roof, 300 yards.

ORDER CRASSULACEÆ.

408. *Sedum Rhodiola*, DC. (Rose-root). Native. Highland type. Range 2-3. Frequent on the higher crags of the slate hills.

C. Piers Ghyll, Mickledore, Red Ore Ghyll, Wastwater Screes, and other cliffs of the Scawfell group; 600-850 yards. First recorded by Mr. Wood. Grassmoor and Grange Fell over Borrowdale.—(Watson.) Ennerdale Coves and Pillar.— (W. Dickinson.) In a gully beneath Glaramara near Stockley Bridge.—(C. Bailey.) Top of Honister Crag.—(W. Foggitt.)

W. Fairfield, and on the Striding-edge cliffs of Helvellyn, up to 900 yards.—(Watson, etc.) Crags south of Easdale Tarn.—(J. Ball.) Blea Water Crag, High Street, and high up in Long Sleddale and Kentmere.—(Wilson.) Mardale.— (Watson.)

L. Coniston Old Man; first recorded by Merrett.

409. *Sedum Telephium*, L. (Great Stonecrop). Native. English type. Range 1. Frequent on hedge-banks and on rocks, both the type and var. *purpureum*, Tausch, ascending from Humphrey Head to 300 yards on Castle Crag, Borrowdale.

410. *Sedum villosum*, L. Native. Highland type. Range 2-3. Damp hills; much rarer in the Lake district than on the Pennine hills.

C. Mosedale, Wastwater.—(W. Dickinson.) Braystones near Egremont.—(J. Robson.) (Is this correct? It seems too low a locality.) Matterdale Common, Ullswater.—(W. Hodgson.) Croglin Fell.—(W. Duckworth.)

W. Swindale, and hillside above the slate quarries in West Sleddale, up to nearly 700 yards.—(Watson.)

L. Clefts of rocks in the north of Furness.—(Aiton.)

412. *Sedum anglicum*, Huds. (White Stonecrop). Native. Atlantic type. Range 1-2. Everywhere common on the slate rocks. St. Bees, Wastwater, Derwentwater, Windermere, Grasmere, Coniston Old Man, Seathwaite Fells, Ulverstone, ascending to Tongue Ghyll waterfall, between Helvellyn and Fairfield, 600 yards. Not seen about Penrith nor round Ullswater by Mr. Hodgson.

Sedum album, L. Alien. One of the commonest of the garden Sedums, well established in many places, as on walls at Ambleside and Milnthorpe, and on the south side of Ullswater about Raven Crag, below Howtown. It also occurs on the conglomerate rocks at the foot of the lake.—(W. Hodgson.)

414. *Sedum acre*, L. (Wall Pepper). Native. British type. Range 1. Rocks, walls, and dry banks. Common; ascending from the sands of the shore in Furness to the limestone cliffs of Shap Common, Whitbarrow, and Farleton Knot, 300 yards, and to Mardale and the Patterdale slate quarries.

Sedum sexangulare, L. Alien. Walls. Rare.

C. Hunday near Workington.—(W. Dickinson.) Watermillock pinfold; introduced.—(W. Hodgson.)

L. Walls at Grange-over-Sands.—(B.)

Sedum reflexum, L. (Love in a Chain, Great Yellow Stonecrop). Alien. Walls and roofs. Not infrequent. Pardshaw Crag, Ullswater, Arnside, Allithwaite, Seathwaite, Great Strickland, etc.

Sedum rupestre, Huds. Alien. Walls. An occasional straggler from gardens. Borrowdale, Windermere near the ferry, Sawrey, Broughton-in-Furness, etc.

Sempervivum tectorum, L. (House Leek, Syphell). Alien. Often seen on roofs and garden walls, as, for instance, at Ritson's Inn at Wastdale Head.

418. *Cotyledon Umbilicus*, L. (Navelwort). Native. Atlantic type. Range 1. Walls and hedge-banks. Rare.

C. Langthatch, Gosforth Bottom, and Ehenside.—(W. Dickinson, J. Robson.) Near Dalston.—(W. Duckworth.) Buckabank.—(Rev. R. Wood.)

W. Troutbeck and other places round Windermere. First recorded by Hudson. In Mireslack, about five miles from Kendal.—(Wilson.)

L. Old wall at Arrad Foot, Ulverstone.—(Miss Hodgson.) Moist rocks near Backbarrow and Ulpha.—(Aiton.)

ORDER SAXIFRAGACEÆ.

Saxifraga Geum, L. Alien. Various Lake stations have been cited for this, and also for *S. umbrosa*, *rotundifolia*, and other cultivated species, but they have no claim to be regarded as Lake plants, except as strays from garden culture. See Borrer, in Phytologist, vol. ii. p. 429, and Watson, Cybele, vol. i. pp. 404-406.

422. *Saxifraga stellaris*, L. (Star Saxifrage). Native. Highland type. Range 1-3. One of the commonest boreal plants of the Lake hills, a great ornament of the high mountain streams.

C. Scawfell Pike from above 1000 feet on the plateau, Helvellyn, Ennerdale hills, Wrynose, Great Gable, Bowfell, Honister Crag, and Glaramara, down to the shores of Derwentwater. Plentiful on Skiddaw and Saddleback.

W. High Street, Kirkstone Pass, Place Fell, Langdale, Easedale, Grisedale, and Glenridding. Plentiful in Long Sleddale about Buckbarrow Well; first recorded by Lawson.

L. Coniston Old Man, Walna Scar, and south of the Three Shire Stones. Not seen about Whitehaven, Penrith, Grange, nor Shap.

423. *Saxifraga nivalis*, L. (Snow Saxifrage). Native. Highland type. Range 3. High slate crags. Very rare.

C. Near the summit of Scawfell.—(J. Robson.)

W. Helvellyn, both on the Striding-edge precipices, at 800-900 yards, and ghylls on the west or Thirlmere slope.— (Balfour, Rev. A. Ley, B.) Cliffs of the High Street range.— (J. Backhouse.)

426. *Saxifraga oppositifolia*, L. (Purple Mountain Saxifrage). Native. Highland type. Range 3. High slate crags. Very rare.

C. Ravine between Scawfell Pikes and Sprinkling Tarn.— (Watson.) Glaramara.—(J. Backhouse.) Ravine of the Wastwater Screes; first recorded by Mr. Wood. Ravine of Great End.—(J. Ball.)

W. Helvellyn, on the Striding-edge cliffs at about 900 yards.—(B.) Found by the Rev. R. Rolleston on rocks above Sweden Bridge. Harrison Stickle, Great Langdale.— (Rev. A. Ley.)

425. *Saxifraga aizoides*, L. Native. Highland type. Range 1-3. Similar in its distribution to *S. stellaris*, and found in similar places.

C. Scawfell Pikes from 800 yards in Piers Ghyll down to the level of Wastwater, Great End, Great Gable, Kirkfell, and Yewbarrow, Haystacks, Grassmoor, Glaramara, and Honister Crag, down Borrowdale, to the foot of Castle Crag. More

ORDER SAXIFRAGACEÆ. 103

sparingly on Skiddaw and the Keswick hills. Rocks in Mosedale just below Floutern Tarn.—(W. B. Waterfall.) Branty Ghyll at the back of Carrock Fell.—(W. Hodgson.)

W. Langdale Pikes, Bowfell, Loughrigg, Helvellyn, and Fairfield, down nearly to the level of Grasmere. High Street and Place Fell, down to Hayes Water. Abundant in Mardale and Long Sleddale.

L. Coniston Old Man, Dobby Shaw, and Cockley Beck Fell, near the copper mines. Not seen about Penrith or in Lower Furness.

427. *Saxifraga granulata*, L. Native. British type. Range 1. Walls and dry banks. Frequent.

C. In the churchyard at Harrington.—(W. Dickinson.) Cliffs near St. Bees Lighthouse, and banks of Keckley Beck. —(E. J. Hughes.) Parton and Egremont, and with double flowers by the Ehen.—(Whitehaven Cat.) Walla Crag, Keswick.—(Winch.) Banks of the Eamont near Yanwath.— (W. Hodgson.)

W. About Maburg and Brougham Castle.—(Mrs. King.) Banks of the Mint near Kendal.—(T. Gough.)

L. Duddon banks and Conishead Priory.—(Aiton.) Dragley Beck wood, Ulverstone.—(Miss Hodgson.)

430. *Saxifraga tridactylites*, L. (Wall Rue). Native. British type. Range 1-3. Walls and rocks. Frequent; ascending from shore level at Kents Bank to the limestone cliffs of Shap Common and Farleton Knot, and to 800 yards on the slate crags of Swirrel Edge.

431. *Saxifraga hypnoides*, L. (Mossy Saxifrage). Native. Scottish type. Range 1-4. Rocky places at all elevations. Frequent.

C. Mickledore, Wastwater Screes, and other crags of the Scawfell group, Armboth Fell, Borrowdale, and the Vale of

Newlands, islands of Ullswater, Honister Crag, Buttermere and Gatescarth Pass, Mockerkin.

W. Helvellyn, Fairfield, High Street, Patterdale, and Kirkstone Pass, abundant, down to the islands of Ullswater. Crags round Easedale Tarn. Mardale and rocks near Crosby Ravensworth. Rocks south of Lowther.—(Lawson.)

L. Coniston Old Man; first gathered by Mr. Jackson. *S. cæspitosa* and *moschata* have both been reported as Lake species, no doubt in error.

434. *Chrysosplenium oppositifolium*, L. (Golden Saxifrage). Native. British type. Range 1-3. Damp woods and mountain streamlets. Common; ascending to the highest springs of Scawfell Pike, Great Gable, and Helvellyn, 850-900 yards.

435. *Chrysosplenium alternifolium*, L. Native. British type. Range 1. In similar places to the other species, but much less common. Windermere, Portinscale, Benson Knot and Hawes Bridge near Kendal, Whitehaven, High Pow near Wigton, Baldhow over Ullswater, Colton Beck Bridge, Duddon Valley, Kirkby-in-Furness, etc.

436. *Parnassia palustris*, L. (Grass of Parnassus). Native. Scottish type. Range 1-2. Everywhere common in swamps, both amongst the slate and limestone hills, ascending to 500 yards at Hayes Water, and Kirkstone Pass, 600 yards.—(Watson.)

ORDER ARALIACEÆ.

437. *Adoxa Moschatellina*, L. (Moschatel). Native. British type. Range 1. Hedge-banks and thickets. Frequent in the lower zone. A two-headed form gathered by Miss Hodgson at Soutergate, Kirkby-in-Furness.

438. *Hedera Helix*, L. (Ivy). Native. British type. Range 1-2. Common on trees, rocks, and hedge-banks through the lower zone, ascending to the limestone pavement of Yewbarrow, Whitbarrow, Farleton Knot, and Shap Common, and to 350 yards in Mardale.—(Watson.)

ORDER CORNACEÆ.

439. *Cornus sanguinea*, L. (Dog-wood). Native. English type. Range 1.

C. Roadside at Gosforth.—(J. Robson.)

W. Not common, but in great beauty round Windermere; probably planted.—(Linton.) Limestone hills south of Kendal.—(Watson.) Hedges at Clawthorpe and near Holme Mill.—(B.) Middlebarrow Wood and other places about Arnside, plentiful and truly wild.—(B.)

ORDER UMBELLIFERÆ.

441. *Hydrocotyle vulgaris*, L. (Penny Wort). Native. British type. Range 1-2. Common in swampy places, ascending from the shore marsh at Arnside to 500 yards at Styhead Tarn.

442. *Sanicula europæa*, L. (Sanicle). Native. British type. Range 1-2. Woods and thickets. Frequent; ascending to the limestone pavement of Farleton Knot and Hutton-Roof, and to 350 yards in Mardale.—(Watson.)

444. *Eryngium maritimum*, L. (Sea Holly). Native. English type. Maritime. Range 1. Sands of the seashore.

C. Common about Whitehaven.—(Whitehaven Cat., W. Hodgson.) Parton near Whitehaven.—(Rev. F. Addison.) Braystones.—(J. Robson.) Seascale.—(T. J. Foggitt.)

L. Island of Walney, and Furness shore at Roosebeck.—(Miss Hodgson.) On the seashore near Quarry Flat, Holker.—(Aiton.)

445. *Cicuta virosa* (Water Hemlock). Native. English type. Range 1.

C. Thursby.—(Rev. R. Wood.)

446. *Conium maculatum*, L. (Hemlock). Native. British type. Range 1. Ditches and roadsides. Frequent; ascending to Shap and Shap Abbey, amongst the ruins, 300 yards, and to Bennet Head over Ullswater.—(W. Hodgson.)

Smyrnium Olusatrum, L. (Alexanders). Alien.

C. Waste ground at Watermillock; introduced.—(W. Hodgson.)

450. *Apium graveolens*, L. (Wild Celery). Native. English type. Maritime. Range 1. Marshes along the coast-line.

C. Marshes at Workington and Kirkbride.—(W. Dickinson.)

W. Arnside and Foulshaw and Brigstear Mosses.—(T. Gough.) Ditches along the shore between Arnside and Milnthorpe.—(B.)

L. About the mouth of Cark Beck, plentiful.—(B.) Plumpton Marsh, and between Old Park and Hag Wood near Holker.—(Aiton, Miss Hodgson.) In the marsh ditches round Cartmel.—(T. Lawson.)

Trinia vulgaris, DC.

C. Tallantire Hill near Cockermouth.—(W. Dickinson.) No doubt a misnomer.

454. *Helosciadium nodiflorum*, Koch. Native. English type. Range 1. Ditches and streams at a low level. Frequent.

Keswick, Ullswater, Kendal, Cartmel, Newby Bridge, Dacre, Stainton, Swarthmore, Townend in the Winster valley, etc.

455. *Helosciadium inundatum*, Koch. Native. British type. Range 1. Ponds and ditches. Frequent.

C. Loweswater.—(W. Dickinson.) St. Bees Moor.— (Whitehaven Cat.) Mockerkin Tarn near Lamplugh.—(W. Hodgson.)

W. Brigstear and other mosses near Kendal.—(T. Gough.) Pond near the Friends' graveyard at Newby Head, where Lawson lies buried.—(B.) Margin of Brotherswater.—(W. Hodgson.) Great Strickland Moor, and at the Roman fort called Maburg, south of Penrith.—(T. Lawson.) Pond near Townend in the Winster valley.—(B.) Crosthwaite, between Kendal and Windermere.—(F. Clowes.)

L. Esthwaite Water and Rusland Moss.—(Mr. Jackson.)

Sison Amomum, L. Alien?

C. Reported by Mr. J. F. Robinson in Exchange Club Report, 1872, p. 26, from the neighbourhood of Penrith.

457. *Ægopodium Podagraria*, L. (Goutweed, Dwarf Elder). Denizen. British type. Range 1. Roadsides and hedge-banks. Not infrequent. Keswick, Lodore, Whitehaven, Coniston, Kirkby Lonsdale, Windermere village, and seen at 250 yards over Penrith, in the road below the beacon.

Carum Carui, L. (Wild Caraway). Alien. Gathered near St. Bees, according to Mr. Adair's Whitehaven Catalogue, and near Shap by Mr. Watson.

459. *Carum verticillatum*, Koch. Native. Atlantic type. Range 1. Gathered on Kingmoor (1882) by my friend, Mr. W. Duckworth of Stanwix, Carlisle.—(W. Hodgson.)

461. *Bunium flexuosum*, With. (Earth Nut). Native. British type. Range 1-2. Pastures and grassy places. Frequent; ascending to Mardale and 1000 feet on Skiddaw; 420 yards.—(Watson.)

462. *Pimpinella Saxifraga*, L. (Burnet Saxifrage). Native. British type. Range 1-2. Pastures and rocky banks. Frequent both on the limestone and slate, ascending to 600 yards on Great End.—(Watson.)

Var. *dissecta* in Ashness Ghyll, Watendlath Valley, etc. *P. magna* is said to have been found at Kidburn Ghyll by the late Mr. W. Dickinson.

Sium latifolium, L.

W. Stock Beck, Kendal.—(T. Gough.) Mr. Gough now thinks a mistake of name has probably been made.

465. *Sium angustifolium*, L. (Water Parsnip). Native. English type. Range 1. Ditches and streams in the low country. Not infrequent. Whitehaven, Ullswater, Kendal, Greystoke, Windermere, Furness, etc.

Bupleurum rotundifolium, L. Alien.

C. Near a corn-mill at Lorton with *Saponaria Vaccaria.*—(W. B. Waterfall.)

470. *Œnanthe fistulosa*, L. Native. English type. Range 1.

C. Ditch near the Old Kiln farm, Allonby.—(W. Hodgson.)

L. Recorded by Wilson as gathered by Lawson close to Grange (and just beyond our limits in ditches between Warton and Carnforth).

471. *Œnanthe Lachenalii*, Gmel. Native. English type. Maritime. Range 1. Coast marshes. Not infrequent.

W. Ditches on the shore between Arnside and Milnthorpe.

L. Salt marsh on Walney Island, nearly opposite Barrow.— (Dr. F. A. Lees.) Plentiful on the shore west of Humphrey Head, and about Cark and Flookborough.—(Miss Hodgson, B.) Salt marsh at Ulverstone.—(Rev. A. Ley.)

473. *Œnanthe crocata*, L. (Dead Tongue). Native. British type. Range 1. Streams and lake-sides in the low country. Common. Derwentwater, Vale of St. John, Windermere, Ullswater, Kendal, Furness, etc. I have seen it at 250 yards near Shap Abbey. 'Kesh,' from which the name of Keswick is said to be derived, is a name given indiscriminately to dried stalks of umbellifers, as Hemlock, Hogweed, Angelica.—(W. Hodgson.)

474. *Œnanthe Phellandrium*, Lam. (Horsebane). Native. English type. Range 1.

C. Bog near Portinscale.—(Winch.) Stream at Grange in Borrowdale.—(J. C. Melvill.) Allonby.—(Mr. Cooke.)

W. Ditches on Brigstear Moss near Kendal.—(Wilson.)

475. *Æthusa Cynapium*, L. (Fool's Parsley). Colonist. British type. Range 1. A frequent weed of cultivated ground. Ascends to 200 yards at Bennet Head, Ullswater.—(W. Hodgson.)

Fœniculum vulgare, Gaertn. (Fennel). Alien.

C. Stray plants at St. Bees, and about the baths at Allonby. —(Whitehaven Cat., W. Hodgson.)

479. *Silaus pratensis*, Bess. (Pepper Saxifrage). Native. English type. Range 1. Pastures and bogs in the lower zone. Not infrequent; ascending to 300 yards.—(Watson.) Whitehaven, Newton Moss, Clibburn, Great Strickland, Dacre, Greystoke, Kendal, Thrimby, etc.

480. *Meum athamanticum*, Jacq. (Bald-money). Native. Scottish type. Range 1.

C. Fell End, Ennerdale. First gathered by Lawson; now extinct. In a field called Bristow Bank near Keswick.—(Otley.)

W. Dunmail Raise.—(Mr. Fardon.) About Shap Wells, 300 yards.—(Watson.) Docker Garths and other places near Kendal.—(Linton.) In a field by the fourth milestone of the road from Kendal to Appleby.—(T. Gough.) At Longwell in Selside, about three miles from Kendal.—(Wilson.) Lambrig Park and Mansergh near Kirkby Lonsdale.—(Hindson.)

L. Furness Fells.—(Mr. Jackson.)

481. *Crithmum maritimum*, L. (Samphire). Native. Atlantic type. Maritime. Range 1. Cliffs of the seashore. Very rare.

C. Cliffs at St. Bees.—(W. Dickinson, W. Hodgson.)

W. Cliffs between Silverdale and Arnside Point.—(C. J. Ashfield.)

L. On Dunnerholme rocks and the cliffs at Humphrey Head. First recorded by Mr. Atkinson, in Withering, edit. iii. vol. ii. p. 295.

482. *Angelica sylvestris*, L. Native. British type. Range 1-2. Swamps and stream-sides. Common; ascending from the shore marshes at Flookborough to 400 yards in Great Langdale, and 560 yards over Sprinkling Tarn.—(Watson.)

Peucedanum palustre, Moench.

L. Ditches at Cannon Winder near Flookborough, and round the sides of Ayside Tarn, three miles north of Cartmel.—(Mr. Hall.) Brought to me and Mr. Crowe when in Lancashire in 1871 by Rev. Mr. Jackson.—(T. J. Woodward.) Modern confirmation wanted.

Peucedanum Ostruthium, Koch. (Masterwort). Alien.

C. Roadside about a mile from Mungrisedale in the direction of Greystoke Park, with *Rumex alpinus*.—(Borrer.) North end of Thirlmere, by a stream in a field on the west side of the road three or four miles from Keswick, with *Polygonum Bistorta*.—(Watson.) One plant near the mouth of Aira Beck, and abundant by the same stream just below Dowthwaite Head.—(W. Hodgson.) Used by cow-doctors in their practice, and, like *Helleborus viridis*, called Fellon-grass.— (W. Hodgson.)

487. *Heracleum Sphondylium*, L. (Cow Parsnip, Hog-weed). Native. British type. Range 1-2. Meadows and pastures. Common; ascending from shore-level in the Furness salt marshes to the limestone pavement of Hutton-Roof, and 500 yards near Sprinkling Tarn.—(Watson.)

Var. *dissecta* in Bowness Woods, and hedges near Allithwaite in Furness.

489. *Daucus Carota*, L. (Wild Carrot). Native. British type. Range 1. Pastures and forage fields in dry soil. Frequent. Abundant about Penrith, Pooley Bridge, and Clibburn. A form with fleshy leaves, which is the *D. maritimus* of Miss Hodgson's Catalogue, is abundant on the Furness shores at Kents Bank, Flookborough, and Plumpton.

Caucalis daucoides, L. Alien.

C. Has appeared as a weed in garden ground at Ghyll Bank, Whitehaven, along with *Camelina sativa* and *Saponaria Vaccaria*.—(W. Hodgson.)

493. *Torilis Anthriscus*, Gaertn. (Hedge Parsley). Native. British type. Range 1. Hedge-banks and amongst rocks. Frequent through the lower zone; ascending to 300 yards near Penrith Beacon, and amongst the limestone cliffs of Shap Common, also at Latrigg, and Castle Crag in Borrowdale.

495. *Torilis nodosa*, Gaertn. Colonist. English type. Range 1. Sandy ground. Very rare.

C. Low Hall, St. Bees.—(Whitehaven Cat.) Sandy ground at Bewaldeth, north of Bassenthwaite Lake.—(W. Dickinson.)

496. *Scandix Pecten*, L. (Shepherd's Needle). Colonist. British type. Range 1. Cultivated ground. Not infrequent; ascending to 250 yards at Tebay.

497. *Anthriscus vulgaris*, Pers. Native. British type. Range 1. Sandy ground. Very rare.

C. Cockermouth and St. Bees.—(Whitehaven Cat.) Sandy soil at Workington Bridge.—(W. Dickinson.) About Goody Hills near Allonby, abundant.—(W. Hodgson.)

498. *Anthriscus sylvestris*, Hoffm. (Wild Chervil, Cow Parsley). Native. British type. Range 1-2. Meadows and hedge-banks. Common in the lower zone, ascending to 350 yards.—(Watson.)

Anthriscus Cerefolium, Hoffm. (Garden Chervil). Alien.

C. In a lane near the church in Patterdale village, 1864.—(W. Foggitt.)

500. *Chærophyllum temulum*, L. (Rough Chervil). Native. British type. Range 1. Hedge-banks and thickets. Frequent in the lower zone; ascending to 300 yards at Shap Abbey. Bullfinches are very fond of the seed.—(W. Hodgson.)

501. *Myrrhis odorata*, Scop. (Sweet Cicely). Denizen. Intermediate type. Range 1. Roadsides and orchards. Not infrequent.

C. Cleator, abundant in several places.—(Rev. F. Addison.) Ormathwaite, but scarcely wild.—(Winch.) About Keswick. (D. Turner.) East side of Derwentwater, looking like a genuine native.—(Watson.) Watendlath.—(W. Foggitt.) In Borrowdale at Seatollar.—(Britten and Holland.) Near the railway station at Drigg.—(J. Robson.) In an orchard at Lodore.—(B.) Banks of the Ehen.—(Linton.) By the Cockermouth road, by a stream a mile and a half south of Lorton.—(B.) Head of Wastwater.—(J. Ball.) Orchards and calf paddocks round Ullswater.—(W. Hodgson.) Banks of streams at Lamplugh and in Ennerdale, to all appearance quite indigenous.—(W. Hodgson, J. Ball.)

W. On the Troutbeck side of Wansfell.—(J. C. Melvill.) Loughrigg Fell.—(J. Ball.) Mardale.—(Watson.) Spital near Kendal.—(T. Gough.) In old orchards round Windermere.—(F. Clowes.)

L. Frequent near old halls and farmhouses in Furness.— (Miss Hodgson.) By the higher road between Grange and Lindale.—(B.) Roadside near Wray Castle.—(B.)

Echinophora spinosa, L. Observed by Mr. Lawson at Roosebeck in Low Furness.—(Ray.) On the shore at Sandside near Ulverstone, and near Winder Hall, Cartmel.—(Aiton.) Modern confirmation specially wanted, as the plant has not been seen in Britain of late years.

ORDER LORANTHACEÆ.

503. *Viscum album*, L. (Mistletoe). Native? English type. Range 1.

W. Very rare in the northern counties, growing only at Lithe near Kendal.—(Mr. Gough : Withering, vol. ii. p. 203, third edition.) Modern confirmation wanted.

ORDER CAPRIFOLIACEÆ.

504. *Sambucus nigra*, L. (Elder). Native. British type. Range 1. Woods and hedges in the lower zone. Frequent.

H

Often planted, but truly wild in the limestone woods about Grange-over-Sands, and on the limestone pavement of Farleton Knot and Hutton-Roof.

505. *Sambucus Ebulus*, L. (Dwarf Elder, Danewort). Denizen. English type. Range 1.

C. Aspatria, in a field on the east of the town.—(Rev. J. Dodd.) St. Bees Valley, rare.—(Whitehaven Cat.) Vale of Lorton.—(W. B. Waterfall.) Brigham near Cockermouth, and at Scalelands and Brackenthwaite.—(W. Dickinson, J. C. Melvill.)

W. A large patch near the farm-house at Glencoin, Ullswater. —(W. Hodgson.) Bradley fields, and Staveley near Kendal. —(T. Gough.) In Troutbeck near the church.—(F. Clowes.)

L. Near Dalton, Bardsea, and Flookborough in Furness.— (Aiton.)

506. *Viburnum Opulus*, L. (Guelder Rose). Native. British type. Range 1. Woods and thickets. Frequent everywhere in the lower zone, and very ornamental both in flower and fruit. Ascends to 250 yards in Naddle Forest, over Hawes Water (Watson); and 300 yards over Coniston.

Viburnum Lantana, L. (Wayfaring Tree). Alien. Parks and shrubberies.

C. Cockermouth.—(Whitehaven Cat.) In shrubberies about Ullswater, introduced.—(W. Hodgson.)

508. *Lonicera Periclymenum*, L. (Honeysuckle). Native. British type. Range 1-2. Woods and hedges. Common in the lower zone, ascending in Great Langdale to Dungeon Ghyll, 500 yards.—(Watson.)

Lonicera Caprifolium, L. Alien.

C. In the Vale of Lorton near Lorton Hall.—(Mr. Tweddle.)

Lonicera Xylosteum, L. Alien.

C. St. Bees.—(J. Robson.) Workington Park.—(Mr. Tweddle.) St. Herbert's Isle, Derwentwater.—(P. F. Lee.)

W. Singleton Park, Kendal.—(T. Gough.) Middlebarrow Wood, Arnside.—(C. Bailey.) Woods in Patterdale.—(Rev. A. Ley.)

L. Head of Coniston Lake near Waterhead.—(B.) In Plumpton Woods and near Conishead Priory.—(Aiton.) Windermere Islands and wood near the Ferry.—(W. Foggitt.)

ORDER RUBIACEÆ.

513. *Galium verum*, L. (Lady's Bedstraw). Native. British type. Range 1-2. Dry banks. Frequent; ascending from the coast sandhills at Maryport, Flookborough, and Bardsea, to the limestone pavement of Farleton Knot and Shap Common, and 370 yards in Mardale.—(Watson.)

514. *Galium cruciatum*, With. (Crosswort). Native. British type. Range 1. Roadsides and hedge-banks. Frequent in the lower zone; ascending from the Furness shore at Flookborough to the limestone cliffs of Shap Common, 300 yards.

515. *Galium palustre*, L. (Marsh Bedstraw). Native. British type. Range 1-2. Swamps and lake-sides. Frequent; ascending to 400 yards in Great Langdale, 500 yards at Hayes Water.

Var. *elongatum* in the moss at Newton Regny near Penrith, etc. Smith records *G. Witheringii* as found by Bishop Goodenough near Rose Castle.

516. *Galium uliginosum*, L. Native. British type. Range 1. Frequent in swamps in the lower zone.

517. *Galium saxatile*, L. (Mountain Bedstraw). Native. British type. Range 1-4. Heaths at all elevations. Ascending to 1000 yards on Skiddaw and Scawfell Pike, 950 yards on Great Gable, 900 yards on Helvellyn, 850 yards on Grisedale Pike.

518. *Galium erectum*, Huds. Native. English type. Range 1.

W. Between Penrith and Pooley Bridge.—(W. Foggitt.)

519. *Galium Mollugo*, L. (Hedge Bedstraw). Native. English type. Range 1. Hedges and thickets. Locally abundant.

C. Crofton Hall, Pardshaw, and Rothersyke near Whitehaven.—(W. Dickinson, Whitehaven Cat.) Valley of St. John.—(B.)

W. Plentiful about Penrith, Pooley Bridge, Clibburn, Great Strickland, Lowther, and Shap, ascending to 300 yards, Kendal.—(T. Gough.) Hedges at Elleray.—(Rev. A. Bloxam.)

L. In the grounds near the railway at Grange-over-Sands. —(B.)

Var. *insubricum*.—Windermere.—(Rev. C. A. Stevens.)

520. *Galium sylvestre*, Poll. Native. Intermediate type. Xerophilous. Range 1-2. Almost confined to the limestone hills.

W. Cliffs of Shap Common, and on the limestone, from Kendal by way of Whitbarrow to Farleton Knot, Hutton-Roof, and Arnside Knot. First recorded by Hudson. Ascending to 450 yards.—(Watson.) Nab Scar near Rydal.—(J. Woods.)

L. On the top of Humphrey Head, with *Asperula cynanchica*.—(Dr. Windsor.) Hampsfield Fell.—(Miss Hodgson.)

523. *Galium Aparine*, L. (Goose Grass; Robin run by t'dyke). Native. British type. Range 1. Hedges and cultivated fields. Common in the lower zone, ascending to 250 yards in Troutbeck Valley; 300 yards.—(Watson.)

G. tricorne, With., is reported from Brookfield by Mr. Cooke. The name requires confirmation.

525. *Galium boreale*, L. Native. Highland type. Range 1-3.

C. Banks of the Borrowdale stream below Grange, and of Derwentwater at Friars Crag, etc.—(W. Dickinson, B.) Near Wastdale Head.—(Mr. Wood.) Piers Ghyll, Scawfell.—(Rev. A. Ley.) Shores of Ullswater at Floshgate and Pooley Bridge. First recorded by Woodward. Cliffs over Keppel Cove Tarn, 800 yards.—(W. Hodgson.) West side of Thirlmere.—(Rev. A. Ley.) On Catbells at 500-550 yards.—(Watson.) Harriman's Penrith station is a mistake for *G. Mollugo*.

W. Shores and islands of Windermere. First recorded by Jackson and Woodward. Drawn from the Ferry by Miss Wilson. In the limestone near Shap.—(T. Gough.) On High Street, over Blea Water.—(Rev. A. Ley.)

L. Shores of Coniston Lake at Waterhead, etc.—(Miss Beever.)

526. *Sherardia arvensis*, L. (Field Madder). Colonist. British type. Range 1. Cultivated fields; ascending to 250 yards over Witherslack Hall; 300 yards.—(Watson.)

527. *Asperula odorata*, L. (Woodruff). Native. British type. Range 1. Woods and thickets in the lower zone. Common; ascending to 300 yards in Mardale, and on the hills between Keswick and Thirlmere.—(Watson.)

528. *Asperula cynanchica*, L. (Quinancy-wort). Native.

English type. Xerophilous. Range 1. Confined to the limestone hills. No record for Cumberland.

W. Beltharrow, and on the top of Cunswick Scar near Kendal, and along the limestone by way of Whitbarrow and Hutton-Roof Crags to Arnside Knot. First recorded by Lawson.

L. Hampsfell and Yewbarrow over Grange, and wood between Grange and Lindale.—(T. J. Foggitt, Miss Hodgson, B.) Plentiful on the top of Humphrey Head, and on the rocks at Copse Head.—(Aiton, Dr. Windsor.)

ORDER VALERIANACEÆ.

Centranthus ruber, DC. (Red Valerian). Alien. An occasional straggler from gardens.

C. Waverton, near houses.—(Whitehaven Cat.)

L. Roadside near Cark station.—(B.)

531. *Valeriana dioica*, L. Native. English type. Range 1-2. Swamps and lake-sides. Frequent; ascending from Newton Regny Moss to upwards of 400 yards on the hills between Keswick and Thirlmere.—(Watson.)

532. *Valeriana officinalis*, L. (Valerian). Native. British type. Range 1-3. Stream-sides and damp places; ascending from shore-level in Low Furness to the limestone pavement of Hutton-Roof Crags, and to 800 yards on slate crags over Sprinkling Tarn.—(Watson.)

Valeriana pyrenaica, L. Alien. An occasional straggler rom gardens.

C. Eskdale.—(J. Robson.)

534. *Valerianella olitoria*, Moench. (Lamb's Lettuce). Colonist. British type. Range 1.

C. Common about Whitehaven.—(Whitehaven Cat.) By

a stream on the sea-coast at Nethertown.—(Rev. F. Addison.) Dry hedge-banks over Ullswater, not common; abundant about Aspatria and Dubmill.—(W. Hodgson.)

W. Frequent about Kirkby Lonsdale.—(Hindson.) About the Roman fort (Maburg) near Penrith.—(Lawson.) Helsington Laythesdale near Kendal.—(J. Wilson.) Drawn from Warcop by Miss Wilson.

L. A form with hairy fruit at Humphrey Head.—(C. Bailey.) Furness shore, not infrequent.—(Miss Hodgson.)

537. *Valerianella dentata*, Koch. Colonist. English type. Range 1. Cultivated fields. Rare.

C. Frizington near Whitehaven.—(W. Dickinson.) In corn-fields at Aspatria.—(Whitehaven Cat.) Corn-fields over Ullswater; almost extinct.—(W. Hodgson.)

W. Arnside Knot.—(D. Oliver.)

L. Furness, in cultivated fields.—(Miss Hodgson.)

ORDER DIPSACEÆ.

539. *Dipsacus sylvestris*, L. (Teazle). Native. British type. Range 1.

C. At Bolton, between Wigton and Cockermouth.—(Rev. R. Wood.)

L. Near Ulverstone, in White Ghyll Wood.—(Aiton, Misses Ashburner.)

541. *Scabiosa succisa*, L. (Hog-a-back; Devil's-bit Scabious). Native. British type. Range 1-3. Grassy places. Very common, ascending to the high slate cliffs. Seen at 700-800 yards on Great Gable, and 800 yards on Scawfell Pike. A white variety on Ulverstone peat-mosses (Miss Hodgson); and a form with a proliferous head seen in the moss below Newton Regny.

542. *Scabiosa columbaria*, L. Native. Xerophilous. English type. Range 1. Cliffs and dry banks. Confined to the limestone.

C. Nethertown near Egremont.—(J. Robson.) Slapestones How, Penrith.—(W. Hodgson.) Hensingham Quarry.—(Whitehaven Cat.)

W. Plentiful at Cunswick Scar near Kendal, and on the limestone hills about Whitbarrow and Arnside; limestone quarries at Clibburn and Great Strickland.—(B.) Closes between Waterfall Bridge and Melkenthorpe near Penrith.—(Lawson.) About Shap and Shap Abbey, up to 350 yards.—(Watson, B.)

L. Rocks of Yewbarrow over Grange.—(W. Foggitt, B.)

543. *Knautia arvensis*, Coult. (Field Scabious). Native. British type. Range 1. Dry banks and cultivated fields. Frequent in the lower zone; ascending to 300 yards in limestone quarries in Shap village, and as high over Coniston by the road up the Old Man.

ORDER COMPOSITÆ.

544. *Tragopogon pratensis*, L. (Goat's Beard). Native. British type. Range 1. Grassy places. Frequent; ascending to 250 yards near Shap.—(Watson.) Abundant on the railway embankment near Corkickle station.

Tragopogon porrifolius, L. (Salsify). Alien.

C. Workington.—(W. Dickinson.) Fields about Rose Castle and Carlisle.—(Nicholson.) Came up on a newly-dug grave in a corner of Aspatria churchyard in 1854. I never saw it near Rose Castle, within half a mile of which I was born.—(W. Hodgson.) Modern confirmation wanted.

546. *Helminthia echioides*, Gaertn. (Ox-tongue.) Native? English type. Range 1.

C. Egremont; perhaps introduced.—(W. Dickinson.)

L. On Oxenfell, north of Coniston.—(Linton.) Confirmation wanted.

547. *Picris hieracioides*, L. Native. English type. Range 1.

C. Bank near the railway station at Keswick.—(B.) Egremont; perhaps introduced.—(W. Dickinson.)

548. *Leontodon hirtus*, L. Native. English type. Range 1. Sandy ground near the coast. Rare.

C. On the coast near Drigg.—(Professor Oliver!) Cleator.—(Rev. F. Addison.) Westward near Wigton.—(Rev. R. Wood.)

L. Humphrey Head.—(Dr. Windsor.)

549. *Leontodon hispidus*, L. (Great Hawkbit). Native. English type. Range 1. Meadows and pastures. Frequent; ascending to 500 yards at Hayes Tarn; 560 yards.—(Watson.)

550. *Leontodon autumnalis*, L. (Lesser Hawkbit). Native. British type. Range 1-3. Grassy places. Common; ascending to 600 yards on High Street and Coniston Old Man, 650 yards at Sprinkling Tarn, 700 yards on Helvellyn.

552. *Hypochœris maculata*, L. Native. English type. Xerophilous. Range 1.

W. Between Kendal and Ambleside.—(Woodward.)

L. Limestone cliffs of the steep west face of Humphrey Head; said to be plentiful by Mr. Hall (Withering, vol. iii. p. 691, third edition), but now nearly or quite extinct.

553. *Hypochœris radicata*, L. (Cat's Ear). Native.

British type. Range 1-2. Grassy places. Frequent in the lower zone; ascending to 400 yards on the hill between Rosthwaite and Watendlath.—(Watson.)

554. *Lactuca virosa*, L. (Wild Lettuce). Native. Germanic type. Range 1.

C. St. Bees.—(Whitehaven Cat.)
L. Very fine on the red sandstone walls of Furness Abbey and in the adjacent quarries.—(W. Foggitt, J. C. Melvill, B.)

557. *Lactuca muralis*, Less. (Rock Lettuce). Native. English type. Range 1. Woods and shaded rocks, both of limestone and slate. Frequent throughout the lower zone; ascending to 300 yards in Mardale (Watson), and Castle Crag in Borrowdale, and to the limestone pavement of Shap Common, Whitbarrow, Farleton Knot, and Hutton-Roof Crags.

Sonchus palustris, L.

W. In some marshy places round Windermere.—(Martineau's Guide.) No doubt a misnomer. A swamp form of *S. arvensis* probably intended.

559. *Sonchus arvensis*, L. (Corn Sowthistle). Colonist. British type. Range 1. A frequent weed of cultivated fields, ascending to 300 yards on Farleton Knot.

560. *Sonchus asper*, Hoffm. (Sowthistle). Native. British type. Range 1. Waste ground. Frequent through the lower zone; ascending to Wastdale Head and 300 yards at Shap and on Farleton Knot.

561. *Sonchus oleraceus*, L. (Sowthistle). Native. British type. Range 1. Waste ground. Less common than the last sub-species; ascending to 300 yards near Shap.—(Watson.)

563. *Crepis virens*, L. (Hawk's Beard). Native. British type. Range 1. Grassy banks. Frequent through the lower zone; ascending to 300 yards over Penrith and Coniston, and 250 yards in Troutbeck Valley.

Var. *agrestis* in cultivated fields at Clibburn and between Hawkshead and Coniston.

Crepis biennis, L.

C. Aspatria.—(Whitehaven Cat.) Mr. W. Nixon, on whose authority the Aspatria station rested, now thinks he was mistaken in its identity.—(W. Hodgson.)

W. Kendal, in pastures on the east side of the valley.— (T. Gough.) Confirmation wanted.

567. *Crepis paludosa*, Moench. (Marsh Hawk's Beard). Native. Scottish type. Range 1-3. Frequent in swamps and damp woods; ascending from nearly shore-level in the moss at Newton Regny to 650 yards near Red Tarn, Helvellyn. —(W. Hodgson.)

568. *Hieracium Pilosella*, L. (Mouse-ear Hawkweed). Native. British type. Range 1-2. Pastures and dry grassy banks. Frequent; ascending to the limestone pavement of Hutton-Roof and Farleton Knot, to 500 yards; in Kirkstone Pass to 570 yards.—(Watson.)

Hieracium aurantiacum, L. Alien. An occasional straggler from gardens.

C. Holmrook Woods near Drigg.—(J. Robson.) On Vicar's Island, Derwentwater.—(Miss Wright, J. B. Davies.) Vale of Newlands, near a cottage, with *Aconitum Lycoctonum*; shown to Borrer by Mr. Wright.

W. Redbank, Grasmere.—(T. J. Foggitt.)

Hieracium Auricula, L.

W. 'Supra Dalehead, non longe a Grassmere in Westmorelandia, sed sparsim.'—(Hudson.) Patterdale.—(Rev. W. Richardson.)

Hieracium dubium, Sm.

W. 'In Monte Fairfield dicta, prope Rydall, in comitatu Westmorelandia.'—(Hudson.) Patterdale.—(Rev. W. Richardson.)

How it was that Hudson came to record these two grossly unlikely species as Lakeland plants I have not been able to make out. They belong to a group quite unlikely to be found anywhere in Britain.

570. *Hieracium alpinum*, L. (*holosericeum*, Backh.). Native. Highland type. Range 3. High slate crags; very rare.

C. Head of Piers Ghyll, Scawfell, and on the rocky summit of Glaramara.—(J. Backhouse, Jr.!)

W. Langdale Pikes.—(Borrer!)

571. *Hieracium chrysanthum*, Backh. Var. *microcephalum*, Backh. Native. Highland type. Range 3. High slate crags. Very rare.

C. Head of Wastdale.—(J. Robson!) Near the summit of Glaramara.—(J. Backhouse, Jr.!)

W. Rocks over Grisedale Tarn.—(J. Flintoft.) On the Striding Edge crags of Helvellyn, and on the High Street range.—(J. Backhouse, Jr.! B.)

572. *Hieracium murorum*, L. (Backh.). Native. British type. Range 1. This sub-species, though not infrequent in Yorkshire, is very rare at the Lakes.

W. Stock Ghyll, Ambleside.—(Borrer.) Westmoreland, probably on the limestone near Kendal.—(Watson!) *H. aggregatum* of the Whitehaven Cat. is no doubt wrongly determined.

572*. *Hieracium cæsium*, Backh. Native. British type. Range 1-2.

C. Piers Ghyll, Scawfell Pike, and on Hobcarten Crag.—(Rev. A. Ley!)

W. Snaka Moss, Ambleside.—(Borrer!) Dollywaggon Pike, Helvellyn.—(Rev. A. Ley!) Rocks in Caudale near the Brothers Water Hotel.—(W. Hodgson!) Walls of Shap Abbey and on the bridge over Hawes Water Beck at Rossgill.—(B.)

573. *Hieracium vulgatum*, Fries. (Common Hawkweed). (*H. sylvaticum*, Smith.) Native. British type. Range 1-2. Cliffs and dry banks at all elevations; ascending from the red sandstone walls of Furness Abbey to the limestone pavement of Farleton Knot and Hutton-Roof, and (var. *maculatum*) to high slate crags of Coniston Old Man; 600 yards. A variety with several stem-leaves approaching *H. tridentatum* grows on the east side of Windermere, below Gunner's How. Smith's *H. maculatum* was founded on a plant brought from Westmoreland by Mr. Crowe.

573*. *Hieracium gothicum*, Fries. Native. Scottish type. Range 1-3.

C. By the side of Derwentwater below Lodore.—(B.) Head of Buttermere Valley below Gatescarth.—(B.) Ghylls of Green Crag, Great End.—(Wright, in herb. Borrer!) Honister Crag.—(Oliver!) Cliffs of Piers Ghyll, Scawfell.—(Rev. A. Ley!)

W. Banks of the river at Colwith Force, and rocks over Easedale Tarn.—(B.) Near Mardale Green, Hawes Water.—(Borrer!) Snaka Moss, on the ridge south of Scandale.—(Borrer!)

575. *Hieracium pallidum*, Biv. Native. Scottish type. Range 1-3. Rocks both of slate and limestone; the most frequent sub-species of the *murorum* group at the Lakes.

C. On the slope of Great Gable over Styhead Pass.—

(Borrer!) Castle Crag, Grange Crag, and other cliffs in Borrowdale.—(B.) Glaramara.—(J. Backhouse, Jr.!) Whiteside.—(Wilson Robinson, in herb. Borrer!) Ennerdale.— (Rev. F. Addison.)

W. Dungeon Ghyll, at the foot of the upper waterfall.— (B.) Stock Ghyll, Rydal Falls, and other places round Ambleside, and rocks near the station at Windermere.— (Borrer!) In Patterdale, on rocks above Hartsop.—(Professor Oliver!) In many places on Helvellyn and Fairfield.— (Woods, etc.) Limestone pavement of Whitbarrow.—(B.) Scout Scar and other limestone cliffs about Kendal.—(T. Gough, etc.) (Considered by Dr. Boswell, in Exchange Club Report, 1879, p. 10, as *H. cinerascens*, Angl. non Jord.) By Buckbarrow Well in Long Sleddale, and rocks by the stream between Anna Well and Shap.—(Lawson.) It is desirable that these last stations should have modern confirmation. I place Lawson's plant here on the authority of Backhouse's monograph. It was from this that the name *Lawsoni* took its origin.

L. Limestone cliffs of Humphrey Head.—(Dr. Windsor.)

575*. *Hieracium argenteum*, Fries. Native. Scottish type. Range 1-2.

C. Slate crags of Great End, and Crags of Mellbreak, over Crummock Lake.—(J. Backhouse.) In Borrowdale, on rocks near the Bowder Stone.—(B.)

W. On the High Street range.—(J. Backhouse.)

576. *Hieracium anglicum*, Fries. Native. Highland type. Range 2-3. High crags, both of slate and limestone. Rare.

C. Cliffs of Great End, over Styhead Pass.—(Borrer!, Oliver!)

W. Rocky ledges of Fairfield at 800-900 yards.—(Rev. H. H. Slater!) Dollywaggon Pike, Helvellyn.—(Rev. A. Ley!) Cliffs of the east face of Helvellyn. First gathered by Woods.

Rocks in Kirkstone Pass due west of the inn (Red Screes).—
(Borrer!) Swindale.—(J. Backhouse, Jr.!) Rocks and ravines
above Mardale Green.—(Watson!) On High Street over Blea
Water.—(Rev. A. Ley!) In Patterdale over Hartsop.—
(Oliver!) Banks of the stream in Stock Ghyll.—(Borrer!)

580. *Hieracium crocatum*, Fries. (including *H. corymbosum*,
Angl.). Native. Highland type. Range 1-2.

C. North shore of Ullswater near Pooley Bridge.—(Borrer!
etc.) Near Hayes Water.—(Borrer!) Banks of the Borrowdale
stream between Grange and the head of Derwentwater.—(B.)
In the Vale of St. John, near the King's Head Inn, near
Thirlmere.—(B.)

W. Great Langdale, and in the Rothay valley near Fox How.
—(Borrer!) Roadside over Hawes Water, near Lord Lonsdale's boathouse.—(B.)

L. River bank, Tongue How, Seathwaite, Duddon Valley.—
(Rev. A. Ley!)

581. *Hieracium strictum*, Fries. Native. Highland type.
Range 1.

C. Vale of Lorton.—(Professor Oliver!)

W. Patterdale.—(J. Backhouse.) On the little knoll near
the church in Mardale, and on the shaded bank of the stream
from Swindale, near Rossgill Bridge, 300 yards.—(Watson!)
H. prenanthoides is given both by Gough and Balfour as a
Lakeland species, but I have never seen a specimen of the
true plant, which occurs sparingly both in Northumberland
and West Yorkshire.

582. *Hieracium tridentatum*, Fries. Native. British type.
Range 1.

W. Banks of the stream that crosses the road in the
Windermere village, below the farm-house south of the railway
station.—(B.)

584. *Hieracium umbellatum*, L. Native. English type. Range 1. Woods and thickets. Rare.

C. Wood at Scotby near Carlisle.—(Professor Oliver!) In Ennerdale, on Kirkland How.—(J. Robson.) Hedge-banks at Drigg.—(Rev. A. Ley.) A drawing in Rooke's Flora, marked 'Bransty Tollgate, Whitehaven (1864).'

W. Woods about Lowther, and on the bank of the Leitch at Common Holm Bridge.—(B.) Roadside at the foot of Hawes Water.—(B.)

L. In Furness near Rampside.—(Professor Oliver!)

584*. *Hieracium boreale*, Fries. (Wood Hawkweed). Native. British type. Range 1-2. Woods and thickets. Common in the lower zone; ascending to 400 yards in Great Langdale; very abundant in the railway cutting south of Ulverstone station.

588. *Taraxacum officinale*, Wigg. (Dandelion). Native. British type. Range 1-3. Waste ground. Common; ascending to 900 yards.—(Watson.)

Var. *palustre* is frequent in swamps, and var. *lævigatum* on dry banks.

590. *Lapsana communis*, L. (Nipplewort). Native. British type. Range 1. Hedge-banks and field-sides. Frequent in the lower zone; ascending to 300 yards over Penrith.

591. *Cichorium Intybus*, L. (Wild Chicory). Colonist. English type. Range 1.

C. Pastures in Ennerdale.—(Whitehaven Cat.) Hensingham near Whitehaven.—(J. Robson.) Sometimes in new-laid pastures, probably introduced among clover or grass seeds.—(W. Hodgson.)

L. Furness, frequent on the borders of fields.—(Aiton.) Near Rampside in Furness.—(Miss Beever.)

592. *Arctium minus*, L. (Burdock). Native. British type. Range 1. Roadsides and waste ground. Frequent in the lower zone; ascending to the top of Whitbarrow, 300 yards, and in Great Langdale to Dungeon Ghyll. The commonest form of the district is *A. intermedium*, Lange. Typical *minus* I have seen at Ulverstone. I have never seen any Lakeland examples of *A. majus*, Schk.

593. *Saussurea alpina*, DC. Native. Highland type. Range 3.

W. Slate crags of Striding Edge, on the west face of Helvellyn, 800-900 yards; first recorded by Woods. Rocks over Keppel Cove Tarn, 800 yards.—(W. Hodgson!)

594. *Serratula tinctoria*, L. (Saw-wort). Native. English type. Range 1.

C. North shore of Ullswater at Waterfoot, etc.; first recorded by Balfour. Woods at Greystoke and in the moss at Newton near Penrith.—(B.) Shores of Bassenthwaite Lake and roadside between Isell and Cockermouth.—(Whitehaven Cat.) Lorton and Embleton.—(W. Dickinson.) Island in Ullswater.—(J. C. Melvill.) Lowes Water.—(W. B. Waterfall.)

W. Sowfield near Great Strickland.—(Lawson.) Barrow-field near Kendal.—(T. Gough.) Banks at Clibburn and Great Strickland.—(B.) About Mardale and Shap, up to 300 yards.—(Watson.)

L. Banks of the river at Newby Head, and on the shore of Windermere at Ferry Inn and in other places.—(B.) Drawn from Wray by Miss Wilson. Islands of Windermere. —(W. Foggitt.) Shore of Coniston Lake at Waterhead.— (Miss Beever.) Frequent in Cartmel and Furness.—(J. Sidebotham.) Hills east of Backbarrow.—(Miss Hodgson.)

595. *Carduus nutans*, L. (Musk Thistle). Native. English type. Range 1-2.

C. In Rooke's drawings, marked 'Kirkland, 1860.'

W. On the limestone about Shap and Kendal, ascending to 370 yards.—(Watson.) First recorded by Lawson.

596. *Carduus crispus*, L. Native. British type. Range 1. Not seen in the heart of the Lake country about Keswick, Ambleside, or Coniston, but abundant about Penrith and Shap, ascending to 300 yards, and seen also at Furness Abbey, Grange-over-Sands, and Carlisle.

597. *Carduus tenuiflorus*, Curt. Native. English type. Maritime. Range 1. Waste ground on the coast. Rare.

C. Egremont and Pardshaw Crag.—(Whitehaven Cat.) In Rooke's drawings, marked 'Beckermont, 1871.'

L. About Barrow in Furness.—(W. Foggitt.) Top of Humphrey Head.—(Miss Hodgson.)

599. *Carduus lanceolatus*, L. (Spear Thistle). Native British type. Range 1-2. Roadsides and waste ground. Common; ascending to the limestone pavement of Whitbarrow and Hutton-Roof, in Borrowdale to the top of Castle Crag, and to 500 yards at Hayes Water.

601. *Carduus palustris*, L. (Marsh Thistle). Native. British type. Range 1-3. Common in damp meadows; ascending to Hutton-Roof, 500 yards on Honister Crag and Styhead Pass, 600 yards on High Street, 660 yards.— (Watson.)

602. *Carduus arvensis*, Curt. (Field Thistle). Native. British type. Range 1-2. Common in cultivated fields and waste grassy places, ascending to the limestone pavement of Farleton Knot, and 500 yards at Hayes Water.

Carduus acaulis, L.
C. Barrowside, Hardknot.—W. Dickinson.) Ennerdale.

—(J. Robson.) In Ennerdale *Carlina vulgaris* grows abundantly, and may have been mistaken for *Card. acaulis*, which I never saw there.—(W. Hodgson.)

W. Dry pastures about Hutton-Roof.—(Hindson.) I am afraid these are all misnomers.

607. *Carduus heterophyllus*, L. (Great Plume Thistle). Native. Scottish type. Range 1. Stream-sides and woods. Not infrequent.

C. Armboth, Watendlath, and hills behind Castlehead, Keswick.—(Watson, etc.) Two places by the road between Seathwaite and Seatollar.—(B.) Banks of Airey stream near Beckbottom Farm, Matterdale.—(W. Hodgson.) Ullock and Papcastle near Cockermouth.—(W. B. Waterfall.) Westward near Wigton.—(Rev. R. Wood.)

W. In Troutbeck Valley in the lane below the highest inn. —(B.) By the roadside near Seathwaite Rayne, Ambleside.— (Mrs. W. H. Hills.) Banks of the stream near the railway junction at Tebay.—(A. W. Bennett.) In Lowther woods near Thrimby.—(B.) In Long Sleddale in a field called Witherhowe, by the roadside near Tomshowe, seven miles from Kendal.—(Wilson.) Between Shap and Orton.—(T. J. Woodward.) Peat Lane, Kendal.—(T. Gough.) Kirkby Lonsdale.— (Hindson.) About Shap, Crosby Ravensworth, Orton and Hawes Water in many places.—(Watson.)

L. Near Newby Bridge.—(Miss Hodgson.)

Onopordum Acanthium, L. Alien?

W. Kirkby Lonsdale.—(Hindson.) Confirmation wanted.

609. *Carlina vulgaris*, L. Native. English type. Range 1-2. Dry hills, both of slate and limestone. Frequent; ascending to the limestone pavement of Shap Common, Whitbarrow, Hutton-Roof, and Farleton Knot, and to 500 yards at Hayes Water.

611. *Centaurea nigra*, L. (Horsenops; Knapweed). Native. British type. Range 1-2. Meadows and pastures. Common; ascending to 350 yards in Great Langdale and Troutbeck Valley. A variety approximating to *C. nigrescens* in Furness by a roadside on the east bank of the river Crayke.—(Miss Hodgson.) A form with long florets about Greystoke.

612. *Centaurea Cyanus*, L. (Corn Bluebottle). Colonist. British type. Range 1. Cultivated fields. Rare.

C. Banks of the Ehen, Egremont.—(Whithaven Cat.) Near Watermillock; scarce.—(W. Hodgson.)

W. Not infrequent about Kirkby Lonsdale.—(Hindson.)

613. *Centaurea Scabiosa*, L. (Great Knapweed). Native. British type. Range 1. Dry banks. Rare.

C. Railway bank at Egremont.—(Whitehaven Cat.) Eaglesfield near Cockermouth.—(W. Dickinson.)

W. About Bampton, 250 yards.—(Watson.) Railway bank near the station at Burton in Kendal.—(B.) Limestone quarry between Clifton and Great Strickland.—(B.) Kendal Fell.—(T. Gough.)

L. Furness shore at Roosebeck.—(Miss Hodgson.) Humphrey Head.—(C. J. Ashfield.) Grange-over-Sands.—(W. Foggitt.)

Centaurea Calcitrapa, L. Alien.

C. Banks of the Ehen at Egremont.—(Whitehaven Cat.) This proved to be *C. melitensis*.

Centaurea melitensis, L. Alien.

C. Introduced with foreign corn at Floshgate, Ullswater, 1882.—(W. Hodgson.)

617. *Bidens cernua*, L. Native. English type. Range 1. Ditches and swamps. Rare.

C. Cloffocks, Whitehaven.—(Mr. Tweddle.) Banks of the stream at Rottington.—(Whitehaven Cat.) Braithwaite near Keswick.—(W. Dickinson.) Sellafield.—(Rev. F. Addison.) Yearn Ghyll pond, Aspatria.—(W. Hodgson.)

W. In a bog near Burneside Hall and Underbarrow toll-bar.—(T. Gough.) Crosthwaite near Kendal.—(F. Clowes.)

L. Peat ditches at Plumpton.—(Miss Hodgson.)

618. *Bidens tripartita*, L. Native. English type. Range 1. Ditches and swamps. Rare.

C. Braystones Tarn and Rottington Beck.—(Whitehaven Cat.) On the shore at Nethertown.—(J. Robson.) Between Seascale and Gosforth.—(W. Foggitt.) Bootle.—(W. Dickinson.) In the Cass at Keswick.—(W. Dickinson.) Northeast of Baggrow near Aspatria.—(Rooke's Flora.)

W. Great Langdale; roadside near the Dungeon Ghyll Inn.—(B.) Stock Beck, Kendal.—(T. Gough.)

Madia racemosa, T. and G. Alien.

C. A Californian annual found with the other casuals by Mr. W. Hodgson at Floshgate, north-east shore of Ullswater.

619. *Eupatorium cannabinum*, L. (Hemp Agrimony). Native. British type. Range 1. Frequent about the Lakes and by stream-sides in the low country. Abundant on the rocky coast from Harrington to Whitehaven.—(W. Hodgson.)

Chrysocoma Linosyris, L. Mentioned in Aspland's list of the rare plants of Grange-over-Sands. Probably a misnomer.

Diotis maritima, Cass. Mentioned in Aspland's list of the rare plants of Grange-over-Sands. Probably a misnomer.

624. *Artemisia maritima*, L. Native. English type. Maritime. Range 1.

C. On the shore at Muncaster, and Coulderton Point.—(J. Robson.) Shore of Duddon estuary.—(W. Hodgson.) Ravenglass.—(W. Dickinson.)

L. Isle of Walney. First recorded by Mr. Atkinson, in Withering, p. 709, 3d edition. Coast between Rampside and Barrow.—(Mr. Gough.)

625. *Artemisia Absinthium*, L. (Wormwood). Denizen. English type. Range 1.

C. About Blencow; once very plentiful, as it was also on Dalston green.—(W. Hodgson.)

W. Abundant on the village green at Askham in 1853.—(T. J. Foggitt.) Bank below the farmhouse near the bridge at Clibburn.—(B.)

L. Lindeth Wood, Rusland.—(Miss Hodgson.)

626. *Artemisia vulgaris*, L. (Mugwort). Native. British type. Range 1. Hedges and field-sides in the low country; not infrequent; ascending to 300 yards at Shap.

627. *Gnaphalium dioicum*, L. (Cat's Foot; Wild Everlasting). Native. Scottish type. Range 1-3. Dry hills and grassy heaths. Frequent; ascending to 650 yards.—(Watson.) Helvellyn, Kendal Fell, Wastdale Screes, Watendlath, Gowbarrow Park, Arnside Knot, Cartmel Fell, Donnerdale Fells, Brockle Crag, north-west side of Skiddaw, abundant, Catlands near Wigton, etc.

Gnaphalium margaritaceum, L. Alien. An occasional straggler from gardens.

C. Near Woodend station.—(Whitehaven Cat.)

W. Roadside between Bowness and Ambleside, 1865.—(J. C. Melvill.)

630. *Gnaphalium sylvaticum*, L. Native. British type. Range 1. Grassy heaths. Not infrequent. Humphrey Head, Clibburn Moor, Whitbarrow, Swinside, etc. Ascends to 300 yards.—(Watson.)

632. *Gnaphalium uliginosum*, L. Native. British type. Range 1. Roadsides and cultivated fields. Frequent; ascending in Buttermere Valley to Gatescarth, and in Great Langdale to Dungeon Ghyll.

Filago gallica, L.

C. Drigg and Gosforth.—(W. Dickinson.) Proved a misnomer for *F. minima*.

634. *Filago minima*, Fries. Native. British type. Range 1. Sandy ground. Rare.

C. Field Head, Eskdale.—(W. Dickinson.) Nethertown near Egremont.—(J. Robson.) Sand-hills south of Maryport.—(Professor Oliver.) Coast sand-hills at Drigg.—(Rev. A. Ley.) West side of Broadfield about Stockdalewath, abundant.—(W. Hodgson.)

L. Foxfield Marsh, and wall-tops both north and south of Ulverstone.—(Miss Hodgson.)

635. *Filago germanica*, L. (Cudweed). Native. British type. Range 1. Sandy soil. Not infrequent; ascending to Wansfell (F. Clowes); and to 300 yards near Penrith Beacon (B.).

636. *Petasites vulgaris*, Desf. (Butter-bur). Native. British type. Range 1. Stream-sides and lake-sides in the low country. Frequent; ascending to 250 yards near Thrimby.

637. *Tussilago Farfara*, L. (Coltsfoot). Native. British type. Range 1-2. Roadsides and earthy banks. Common; ascending to 600 yards.—(Watson.) The highest station in which I have noted it is about the slate quarries on Coniston Old Man. It ascends to the limestone pavement of Farleton Knot.

639. *Erigeron acris*, L. (Fleabane). Native. English type. Range 1. Grassy banks. Very rare.

C. Gravelly bank of the Caldew at Dalston, sparingly.—(W. Hodgson, Rev. R. Wood.)

W. Foulshaw Moss, Kendal.—(T. Gough.) Arnside.—(Miss Beever.)

Erigeron canadensis, L. Alien.

L. On Yewbarrow, over Grange-over-Sands, 1872.—(Hindson.)

641. *Aster Tripolium*, L. (Starwort). Native. British type. Maritime. Range 1. Salt marshes. Common.

C. Little Mill reservoir, Workington Marsh, and Black Beck.—(Whitehaven Cat.) Ravenglass.—(J. Robson.)

W. Abundant about Arnside.

L. Abundant all along the Morecambe estuary from Walney Island and Bardsea, past Cark and Flookborough to Grange. First recorded by Lawson.

Cultivated American Asters (Michaelmas Daisy), either *Aster lævis* or allied species, are established on the shores of Ullswater, Windermere, and Derwentwater.

642. *Solidago Virgaurea*, L. (Golden Rod). Native. British type. Range 1-3. Everywhere frequent in woods and on rocks, ascending to the high slate crags; 800 yards over Sprinkling Tarn, and on Helvellyn and Scawfell Pike; 600 yards on Great Gable and Coniston Old Man.

Var. *cambrica*, Huds., is the ordinary form on the high cliffs.

643. *Senecio vulgaris*, L. (Groundsel). Native. British type. Range 1-2. Cultivated and waste ground. Common; ascending to 300 yards at Shap; 350 yards.—(Watson.)

644. *Senecio sylvaticus*, L. Native. British type. Range 1-2. Sandy moors. Frequent; ascending from the shore at Meathop Moss to Priest Crag, Ullswater; 350 yards on Robinson, and 300 yards on Latrigg.

645. *Senecio viscosus*, L. Native. Germanic type. Range 1.

C. On the coast at Nethertown.—(J. Robson.) Parton near Whitehaven.—(Rev. F. Addison.) Cleator Moor.—(Whitehaven Cat.) Common between the railway station and the sea at Workington.—(T. J. Foggitt, W. Hodgson.)

L. Sandy bank in Walney Island near the Ferry.—(Dr. F. A. Lees.)

647. *Senecio erucifolius*, L. Native. English type. Range 1.

C. At Little Broughton, between Cockermouth and Workington.—(W. Dickinson.) It has been found by Mr. W. Duckworth beyond our limits at Longtown, north of Carlisle.

648. *Senecio Jacobæa*, L. (Ragwort; 'Booins'). Native. British type. Range 1-2. Grassy places. Common in the lower zone; ascending to 360 yards on Skiddaw (Watson); and about as high on the slope of Great Dodd to the Vale of St. John. A discoid variety by the side of the higher road between Grange and Lindale.

649. *Senecio aquaticus*, Huds. (Marsh Ragwort). Native.

British type. Range 1-2. Stream-sides and swamps. Common; ascending from the shore marsh below Middlebarrow Wood, Arnside, to 400 yards in Hag Ghyll, Troutbeck; 500 yards.—(Watson.)

650. *Senecio saracenicus*, L. Denizen. Intermediate type. Range 1.

C. Hedges near Longtown and Eden side below Carlisle. —(Mr. Jackson.) Near Hutton Hall, north of Penrith.— (Borrer.) It grows as plentifully in the fields at Salkeld as the *vulgaris* (i.e. *Jacobæa*).—(Mr. Nicholson, Ray, Syn. iii. 177.) Houray pasture, Keswick.—(J. Otley.) At Moresby near Whitehaven.—(J. Robson.) Sebergham.—(W. Dickinson.) Paddock at Thackthwaite, and near the old homesteads at Stoddah.—(W. Hodgson.)

W. Stock Beck, Kendal.—(T. Gough.)

L. Roadside at Newby Bridge, and in an old orchard at Ghyll Head; first recorded by Woods. Corner of a field on the Old Hall estate, Ulverstone.—(Miss Hodgson.)

Cineraria campestris, Retz.

C. Cold Fell, Egremont.—(J. Robson.) I fear a misnomer, but it has been found lately by Backhouse on Micklefell, an unexpected extension of range.

Cineraria palustris, L. A plant recorded by Wilson from Burton Moss, Westmoreland, and Pillin Moss, Lancashire, has been referred to this species. Most likely a misnomer.

Doronicum Pardalianches, L. (Leopard's Bane). Alien.

C. West side of the fish-pond at Brayton Hall, introduced. —(W. Hodgson.) Kelton and Isell near Cockermouth.— (W. Dickinson.)

W. Hedge-bank at the foot of a garden on the high-road between Ambleside and Rydal, 1864.—(C. Bailey.)

655. *Inula Helenium*, L. (Elecampane). Denizen. English type. Range 1.

C. At Mosser near Lowes Water.—(W. Dickinson.)—Coulderton Point.—(J. Robson.) Little Mill, Egremont; probably a relic of cultivation.—(Whitehaven Cat.)

W. About farmhouses near Cunswick.—(T. Gough.)

L. Several places near Dalton in Furness, clearly wild.—(Mr. Atkinson, in Withering, vol. iii. p. 731, 3d edition; Aiton.)

656. *Inula Conyza*, DC. (Ploughman's Spikenard). Native. Xerophilous. English type. Range 1.

W. Cunswick Scar, and abundant on Whitbarrow, and along the limestone to Farleton Knot and Arnside Knot. First recorded by Wilson.

L. Humphrey Head, Kents Bank, Yewbarrow, and other places on the limestone about Grange and Cartmel. First recorded by Lawson.

658. *Pulicaria dysenterica* (Gaertn.). Native. English type. Range 1. Damp places. Very rare.

C. St. Bees Head.—(J. Robson.) Etterby Scar, Carlisle.—(W. Duckworth.)

W. Barrowfield Wood, Kendal.—(T. Gough.)

660. *Bellis perennis*, L. (Daisy). Native. British type. Range 1-3. Meadows and pastures. Common; ascending to 500 yards at Hayes Water and Kirkstone Pass; 700 yards.—(Watson.)

661. *Chrysanthemum segetum*, L. (Corn Marigold). Colonist. British type. Range 1. Cultivated fields. Not infrequent. Nethertown, Duddon Valley, Penrith, Warcop, Ennerdale, Egremont, Newton, Colwith, Hawkshead, Bowness, Grange-over-Sands, Ulverstone, etc.

662. *Chrysanthemum Leucanthemum*, L. (Ox-eye Daisy). Native. British type. Range 1-2. Meadows and pastures. Common in the lower zone; ascending to 350 yards.—(Watson.)

Pyrethrum Parthenium, Sm. (Feverfew). Alien. A frequent straggler from village and farmhouse gardens. Furness Abbey, Arnside Tower, Stonethwaite, Lorton, Bowness, etc.

663*. *Tanacetum vulgare*, L. (Tansy). Denizen. British type. Range 1.

C. Ellercar and Tallantire near Maryport.—(W. Dickinson.) Roadside at Gutherscale, Derwentwater, at the foot of Catbells.—(C. Bailey.) Ullswater; not common, and probably a garden outcast.—(W. Hodgson.) Middletown, Caldew, and Ennerdale, always near houses.—(Whitehaven Cat.) Roadside near Stainton.—(B.)

W. Not infrequent about Kirkby Lonsdale.—(Hindson.) Roadside at Clawthorpe.—(B.) Bank beneath the farmhouse nearest the bridge at Clibburn.—(B.)

L. Bardsea, doubtfully wild.—(Miss Hodgson.)

664. *Pyrethrum inodorum*, Sm. Native. British type. Range 1. Waste ground and cultivated fields. Common; ascending to 300 yards at Tebay, Shap, and Penrith. A maritime variety is abundant along the Furness shore from Walney Island, past Cark and Flookborough, to Arnside, and along the Cumbrian shore as far north as Maryport.

665. *Matricaria Chamomilla*, L. (Wild Camomile). Colonist. English type. Range 1. Waste ground. Very rare.

C. Sylcroft.—(W. Dickinson.) Roadsides at Beckermet.—(Whitehaven Cat.) Introduced with foreign corn at Flosh gate, Ullswater.—(W. Hodgson.) West Newton and New Cooper near Allonby, about stackyards.—(W. Hodgson.)

668. *Anthemis arvensis*, L. (Corn Camomile). Colonist. English type. Range 1. Forage fields. Not infrequent; ascending Borrowdale to Grange and the Vale of St. John to the foot of Thirlmere. By *Anthemis maritima*, in Robson's list, recorded from the shore at Coulderton, *Pyrethrum* is no doubt intended, as it grows there.

669. *Anthemis Cotula*, L. Colonist. English type. Range 1. Cultivated fields. Very rare. I have seen it once only, near Penrith, and it is recorded by Dickinson from a roadside between Seascale and Calder Bridge, and Mr. Roper gathered it near Skelwith Force.

Anthemis nobilis, L. (Camomile). Alien. Waste ground beside the farmhouse nearest Colwith Force, 1882.—(B.)

670. *Achillea Ptarmica*, L. (Sneezewort). Native. British type. Range 1-2. Frequent in damp pastures, ascending to 350 yards.—(Watson.)

671. *Achillea Millefolium*, L. (Yarrow; Milfoil). Native. British type. Range 1-3. Common in grassy places, ascending to 600 yards on Great Gable; 800 yards.—(Watson.)

ORDER CAMPANULACEÆ.

675. *Campanula rotundifolia*, L. (Harebell). Native. British type. Range 1-4. Heaths and pastures. Frequent at all elevations; ascending to 1000 yards on Helvellyn and Skiddaw.

Campanula rapunculoides, L. Alien.
C. Near Moor Row Station.—(Whitehaven Cat.)
W. Roadside at Clappersgate, 1882.—(B.)

678. *Campanula latifolia*, L. (White Foxglove; Giant Bellflower). Native. Scottish type. Range 1-2. Woods and hedges. Frequent. Probably nowhere in Britain does the great bell-flower grow in greater plenty than at the Lakes, as for instance in Troutbeck Valley, and about Lowther, Keswick, and Pooley Bridge. Mr. Watson notes it at 350 yards in Mardale, and it ascends nearly or quite as high near Watendlath. Wilson says that at Kendal they sometimes strip the skin off the young stems and eat them like asparagus.

Campanula Trachelium, L.

W. In the hedges a little way from the footpath leading from Sizergh to Levens.—(Wilson.) Modern confirmation wanted. The Keswick plant given by Linton as *C. Trachelium* is the last species.

681. *Campanula glomerata*, L. Native. Germanic type. Range 1. Confined to the limestone tract about Shap and Kendal. First recorded by Lawson. Crosby Ravensworth, Askham, Pooley Bridge, Great Strickland, and on the Cumberland side of the Eamont at Redhills, and also plentiful about Penruddock.

687. *Jasione montana*, L. (Sheep's Scabious). Native. British type. Range 1-2. Wall-tops, roadsides, and dry banks. Frequent. Buttermere, Grasmere, Thirlmere, Coniston Water, Hawes Water, Windermere, Ennerdale, etc., ascending to 400 yards in Great Langdale. It is a rare plant in the north-eastern counties.

689. *Lobelia Dortmanna*, L. Native. Scottish type. Range 1-2. In all the larger lakes and several of the tarns. Windermere, Grasmere, Derwentwater, Ullswater, Thirlmere, Ennerdale Lake, Rydal Water, Coniston Water, Little Langdale Tarn, Floutern Tarn, Watendlath Lower Tarn, and Blea

Tarn (500 yards), etc. Not known anywhere in the eastern counties of the north of England, but plentiful in Wales. The same is the case with its frequent associate *Isoetes*.

ORDER ERICACEÆ.

690. *Erica Tetralix*, L. (Cross-leaved Heath). Native. British type. Range 1-2. Common on the damper moors; ascending from the shore-level at Meathop and Foulshaw as high up as *Pteris*, but, like *E. cinerea*, falling far short of the *Calluna*. The highest place where I have noted it is 500 yards over Gatescarth Pass. Mr. Watson saw it at 550 yards. Mr. Bailey found in Borrowdale, in June 1865, a monstrosity with free petals.

692. *Erica cinerea*, L. (Fine-leaved Heath). Native. British type. Range 1-2. Common on the drier hills; ascending to 550 yards on Great Gable.

695. *Calluna vulgaris*, Salisb. (Ling). Native. British type. Range 1-3. Everywhere the common heather of the moors; ascending from the shore peat-mosses at Meathop to 900-1000 yards on Skiddaw.

699. *Andromeda polifolia*, L. Native. Intermediate type. Range 1.

C. Moss at Broomfield near Allonby.—(Rev. J. Dodd.) Drumburgh and at Moresby near Whitehaven.—(W. Dickinson, J. Robson.) Moor near Latrigg.—(R. Calvert.) Wedholme Flow near Wigton.—(W. B. Waterfall.)

W. Brigstear Moss, Kendal; first recorded by Lawson. Heversham Moss, Milnthorpe.—(Hindson.) Plentiful in the mosses at Ulpha and Witherslack.—(Rev. A. Ley.)

L. Causeway Moss and Rusland Moss, Furness Fells.—
(Mr. Jackson.) Ellerside Moss.—(Aiton.) Angerton Moss,
Ireland Moss, Roam Moss, Stockbird Moss; on all the good
peat tracts between Ulverstone and Haverthwaite.—(Miss
Hodgson.)

Arbutus alpina, L.

C. Scawfell.—(J. Robson.) No doubt a misnomer.

701. *Arbutus Uva-Ursi*, L. (Bearberry). Native. Highland type. Range 2.

C. On Corney Fell near Bootle.—(J. Robson.) On the slope of Grassmoor towards Crummock Water, at about 500 yards.—(Watson.) Brackenthwaite.—(Wilson Robinson.)

W. Dale-head in Martindale.—(Rev. W. Richardson.)

703. *Vaccinium Myrtillus*, L. (Bilberry; Bleaberry). Native. British type. Range 1-4. Woods and heaths at all elevations; ascending from shore-level at Meathop Moss to 1050 yards on Helvellyn, and also to the summits of Scawfell Pike, Skiddaw, Great Gable, Grassmoor, and Grisedale Pike. Very fine in the fir woods of Penrith Beacon, covering acres of ground.

704. *Vaccinium uliginosum*, L. Native. Highland type. Range 1-2.

C. Moorside Park near Lamplugh.—(W. Dickinson.) Mill Fell.—(Winch.) Near Armathwaite Castle.—(Rev. R. Wood.)

W. Whinfell Forest near Penrith.—(T. Lawson.) On the High Street range.—(J. Backhouse.)

L. Peat mosses at Holker.—(Aiton.)

705. *Vaccinium Vitis-Idæa*, L. (Whortleberry; Cowberry). Native. Highland type. Range 2-4. On all the higher hills, but not nearly so plentiful as the Bilberry; ascending to

1000 yards on Helvellyn, and almost to the summit of Skiddaw, Great Gable, Pillar, Grassmoor, and Grisedale Pike.

706. *Vaccinium Oxycoccos*, L. (Cranberry, Crones, and Cranes). Native. British type. Range 1-2. Damp heaths. Not infrequent.

C. Ennerdale, Black Moss, etc.—(Rev. F. Addison, Whitehaven Cat.) Floutern Tarn.—(W. Foggitt.) Swamps near the highest point of the road over Whinlatter.—(Watson.) Mockerkin.—(W. B. Waterfall.) Broomfield near Allonby.— (Rev. J. Dodd.) Hutchinson says a bog near Hesket-in-the-Forest has yielded £20 worth of cranberries in a year. This bog (Tarn Wadling) has since been drained.—(W. Hodgson.) South side of Skiddaw near the foot of the hill.—(Watson.) Mosedale.—(Rev. A. Ley.)

W. Whinfell Forest near Penrith.—(T. Lawson.) Frequent on boggy moors round Kendal.—(Curtis.) Mardale (Watson); 450 yards. Swamps in Easedale.—(B.) In a few places round Windermere.—(F. Clowes.) Frequent round Ullswater.—(W. Hodgson.) Ridge between Fusedale and Fordendale.—(W. Foggitt.) Bog near Orrest Head.—(F. C. Roper.) Ulpha Moss.—(Rev. A. Ley.)

L. High marshy ground over Coniston.—(Miss Beever.) Kirkby Moor, Furness.—(Miss Hodgson.) Stribers Moss, Haverthwaite.—(Aiton.)

708. *Pyrola media*, Sw. (Winter Green). Native. Scottish type. Range 1.

C. Brayton Woods.—(Whitehaven Cat.) Kirklington Moors.—(W. Dickinson.) With *P. secunda* below the precipitous part of Walla Crag.—(J. Woods.)

W. Stock Ghyll Force, on rocks and banks near the waterfall. —(Watson, etc.)

This has been confused both with *rotundifolia* and *minor* in the Guide-books.

709. *Pyrola minor*, L. (Lesser Winter Green). Native. Scottish type. Range 1.

C. In plenty in a small wood below Penrith Beacon, where it was shown to me in 1883 by Mrs. King. Flimby Wood near Maryport.—(W. Hodgson.) Wigton.—(W. B. Waterfall.) In Rooke's drawings, 'Whitecroft Wood, Tallantire, 1842.'

710. *Pyrola secunda*, L. Native. Scottish type. Range 2.

C. In Ashness Ghyll above Barrow Force, where it was shown to Winch and Dawson Turner by Hutton. Extinct in 1835. —(Woods.) With *P. media* in the wood at Walla Crag.— (Woods, W. Matthews.) In a ravine between Great Dodd and Helvellyn, over King's Head Inn, Thirlmere, at about 500 yards.—(Woods, Watson, Ley.) Westward near Wigton. —(Rev. R. Wood.)

Pyrola uniflora, L. Reported by the late Mr. Wright of Keswick from a wood at Bardsea. No doubt a misnomer.

712. *Monotropa Hypopitys*, L. (Bird's Nest). Native. Germanic type. Range 1. Shaded woods. Very rare.

W. Barrowfield Wood, Kendal.—(T. Gough.) Holme area, below Kirkby Lonsdale Bridge; two plants seen in 1836.—(Hindson.)

ORDER ILICACEÆ.

713. *Ilex Aquifolium*, L. (Holly). Native. British type. Range 1-2. Common in woods and hedges everywhere in the lower zone; ascending to the limestone pavement of Whitbarrow and Hutton-Roof, to 400 yards in Great Langdale and Troutbeck Valley, to 500 yards on Catbells, over Derwentwater.—(Watson.)

ORDER JASMINACEÆ.

714. *Ligustrum vulgare*, L. (Privet). Native. English type. Range 1. Truly wild on the limestone cliffs, as on Arnside Knot and the steep western escarpment of Humphrey Head; ascending with Yew to the limestone pavement of the top of Whitbarrow; also often planted in hedges, as at Ambleside and Bowness, in the Vale of St. John, at the foot of Ullswater, and at Whitefield House, Overwater.

715. *Fraxinus excelsior*, L. (Ash). Native. British type. Range 1. One of the commonest trees of the Lake woods; ascending to 400 yards in Great Langdale, 350 yards on Grisedale Pike and the hills between Rosthwaite and Watendlath. The finest tree I remember stands just within the entrance gate of the grounds of Furness Abbey.

ORDER APOCYNACEÆ.

Vinca minor, L. (Periwinkle). Alien. Frequent in parks and near gardens, as for instance at Coniston Water-head and just north of Newby Bridge.

ORDER GENTIANACEÆ.

Gentiana verna, L.

C. Till lately on Egremont Green.—(W. Dickinson.)

L. Has been found amongst the hills in the north of Furness.—(Aiton.) Probably a misnomer in both instances.

719. *Gentiana Pneumonanthe*, L. (Heath Gentian). Native. English type. Range 1. Damp heaths. Very rare; probably now extinct.

C. Field between Maryport and Flimby; 200-300 yards from the latter.—(Rev. J. Harriman.) Now extinct.

W. Foulshaw Moss near Milnthorpe. First recorded by Hudson.

L. Rosecote near Dalton in Furness.—(Aiton.) Walney Island.—(Rev. Mr. Jackson.)

721. *Gentiana Amarella*, L. (Common Gentian). Native. British type. Range 1. Dry banks. Not infrequent, especially on the limestone, ascending from near shore-level at Grange-over-Sands to the summit of Whitbarrow, and 300 yards at Shap.

722. *Gentiana campestris*, L. Native. British type. Range 1. In similar places to *G. Amarella*, with which it is often associated. Ascending from shore-level at Braystones, Workington warren, and Maryport, to 300 yards in Mardale. —(Watson.)

724. *Erythræa Centaurium*, Pers. (Centaury). Native. British type. Range 1. Pastures and dry banks. Frequent; ascending to the top of Hampsfell, over Grange-over-Sands. The *E. latifolia*, recorded by Woods from the coast at Ravenglass, is probably a variety of this.

724*. *Erythræa littoralis*, Pers. Native. British type. Maritime. Range 1.

C. At Braystones near Egremont.—(J. Robson.) Between Seascale and Gosforth.—(T. J. Foggitt.) Gathered also by Harriman, probably near Maryport.

W. Edge of Ulpha Moss, and on the opposite shore of the Morecambe estuary under Arnside Knot, with *E. pulchella*.—(C. Bailey, B.)

L. Plumpton salt-marshes, and also on the shore under

Humphrey Head.—(Miss Hodgson.) High Cartmel medicinal well.—(Ray, as a var. of *Centaurium.*)

724*. *Erythræa pulchella*, Fries. Native. English type. Range 1.

W. With *E. littoralis* at Ulpha and Arnside.—(C. Bailey, B.)

L. Pasture near Greenhills, Plumpton.—(Miss Hodgson.)

Villarsia nymphæoides, Vent. Alien.

L. In the Moss river near Hawkshead. First recorded by Lawson. I could not find it there in 1883.

727. *Menyanthes trifoliata*, L. (Bog Bean). Native. British type. Range 1-2. Swamps and lake-sides. Frequent; ascending to 600 yards on Haystacks, and on the hills between Keswick and Thirlmere.—(Watson.) Very fine in the two reservoirs at Holme Mill, with *Hippuris.*

ORDER POLEMONIACEÆ.

Polemonium cæruleum, L. (Jacob's Ladder). Alien?

C. St. Bees.—(J. Robson.) By the river Ellen near Torpenhow.—(W. Hodgson.)

W. At Kendal, on the east side of the river Kent, between the mill-race and Kirk-dub.—(Wilson.) Near the limekilns at Kendal, scarce.—(T. Gough.)

L. Graythwaite woods, west side of Windermere.—(F. Clowes.) Reported from Furness Fells by Ray, on the authority of Willisel. Truly wild in the Westmoreland portion of Teesdale.

ORDER CONVOLVULACEÆ.

729. *Convolvulus arvensis*, L. (Bindweed). Native. English type. Range 1. Not seen in the heart of the Lake

country about Keswick and Ambleside or Coniston, but occurring sparingly on the outskirts at Whitehaven, Ulverstone, and Kirkby Lonsdale. Field by river Ellen, below Blennerhasset.—(W. Hodgson.)

730. *Convolvulus sepium*, L. (Great Bindweed). Native. English type. Range 1. Woods and hedges. Frequent; ascending Borrowdale to Seathwaite, Great Langdale to Dungeon Ghyll Inn, and to 250 yards on Brantfell over Bowness. There is a beautiful pink-flowered variety, and both this and the type are often grown in the cottage gardens and trained round porches, like *Ipomæa* in the south of England. Mr. W. Hodgson says of it, 'In some parts of Cumberland sadly too plentiful.'

731. *Convolvulus Soldanella*, L. (Sea Bindweed). Native. English type. Maritime. Range 1. Coast sand-hills. Rare.

C. Coulderton, Parton, and Harrington.—(W. Dickinson, etc.) Maryport.—(Winch.) Lowea.—(Rev. F. Addison.) Abundant at Seascale.—(W. Foggitt.) Braystones.—(J. Robson.) St. Bees.—(W. B. Waterfall.)

L. Walney Island. First recorded by Lawson; also by Dalton and Aiton.

Cuscuta europæa, L.

C. Said to have been found by W. Dickinson at Graysouthen near Workington. Confirmation wanted.

ORDER SOLANACEÆ.

736. *Hyoscyamus niger*, L. (Henbane). Native. English type. Range 1. Waste ground, especially near the coast; not persistent in its localities. Cockermouth, Flimby, St. Bees, Harrington, Allonby, Walney Island, Bardsea, Grange-over-Sands, Levens, Arnside, etc.

Solanum nigrum, L. Alien?

C. Has occurred as a garden weed at Stanwix near Carlisle,
—(W. Duckworth.)

L. Abundant at Barrow in Furness.—(W. Foggitt.)

738. *Solanum Dulcamara*, L. (Bitter-sweet). Native. British type. Range 1. Hedges and thickets. Not infrequent in the low country, ascending to 200 yards over Hawes Water. —(Watson.) A maritime variety at Drylands, Isle of Walney. —(Miss Hodgson.)

739. *Atropa Belladonna*, L. (Deadly Nightshade). Native. English type. Range 1.

C. Once plentiful near Egremont Castle, but now extinct.— (W. Dickinson.) Isell Hall Woods, Cockermouth.—(Rooke's drawings.)

W. In Curren Wood kins near Burton in Kendal.— (Wilson.) Middlebarrow Wood and Hag Wood, Arnside.— (G. S. Gibson, B.)

L. Amongst the ruins of Furness Abbey; first recorded by Mr. Atkinson (Withering, p. 253, 3d edition). Hedge on the shore west of Humphrey Head.—(B.) Conishead Bank, Copse Head, and on the shore near Canon Winder.— (Aiton.) In the park at Holker.—(Rev. A. Ley.)

ORDER SCROPHULARIACEÆ.

740. *Verbascum Thapsus*, L. (Shepherd's Staff; Great Mullein). Native. English type. Range 1. Dry banks. Frequent, especially on the limestone, ascending from shore-level at St. Bees and Coulderton to the limestone pavement of Farleton Knot and Whitbarrow.

Verbascum nigrum, L., is said to have been found by the Rev. R. Wood at Westward near Wigton.

Verbascum Blattaria, L. Alien.

C. Churchyard at Aspatria.—(Rev. J. Dodd.) Appeared there by scores in the Vicarage grounds after alterations in 1872.—(W. Hodgson.)

Verbascum virgatum, With. Alien.

L. In Furness at Roosebeck.—(R. Ashburner.)

746. *Veronica spicata*, L. Native. English type. Range 1. Cliffs and dry banks. Very rare.

W. Arnside Park.—(E. Robson.)
L. Limestone rocks of Humphrey Head, in considerable quantity, at the south end of the promontory. First recorded by Mr. Hall, in third edition of Withering. Penny Bridge, north of Ulverstone.—(T. J. Woodward.) Gathered by Miss Beever in Silverdale, just beyond our limits.

747. *Veronica arvensis*, L. Native. British type. Range 1. Cultivated fields and waste ground. Frequent; ascending to 300 yards on Latrigg (B.); 500 yards (Watson).

750. *Veronica serpyllifolia*, L. Native. British type. Range 1-3. Roadsides and pastures. Frequent; ascending to 500 yards.—(Watson.)

Var. *humifusa*, Dicks., on the slate cliffs of the east face of Helvellyn, 800-900 yards.—(Balfour.)

753. *Veronica scutellata*, L. Native. British type. Range 1-2. Frequent in swamps, ascending to 500 yards on the Styhead Pass.

754. *Veronica Anagallis*, L. (Water Speedwell). Native. British type. Range 1. Ditches and slow streams. Frequent in the lower zone; ascending from the shore marshes

ORDER SCROPHULARIACEÆ.

at Allonby, Arnside, Cark, and Flookborough, to 300 yards at Shap.

755. *Veronica Beccabunga*, L. (Brooklime). Native. British type. Range 1-2. Streams and ditches. Common; ascending from shore-level at Bardsea to 400 yards in Hag Ghyll, Troutbeck; 510 yards.—(Watson.)

756. *Veronica officinalis*, L. Native. British type. Range 1-2. Heathy pastures. Frequent; ascending to the limestone pavement of Whitbarrow, 500 yards on Coniston Old Man near Low Water; 560 yards.—(Watson.)

757. *Veronica montana*, L. Native. British type. Range 1. Not infrequent in shady woods.

C. Snebra near Whitehaven.—(Whitehaven Cat.) Roadside between Keswick and Lodore.—(Watson.) Moor Row near Whitehaven.—(Rev. F. Addison.) Walla Crag Wood, Keswick.—(Winch.) Dunmallet and Airey Woods, Ullswater. —(W. Hodgson.)

W. Middlebarrow Wood, Arnside.—(B.) At Buckham near Lowther.—(T. Lawson.) Fox Ghyll, Ambleside, drawn by Miss Wilson.

L. Miller Bay, Windermere.—(Rev. A. Bloxam.) Islands of Windermere.—(W. Foggitt.) About Coniston.—(Miss Beever.)

758. *Veronica Chamædrys*, L. (Germander Speedwell). Native. British type. Range 1-2. Grassy places. Common; ascending to the top of Hutton-Roof, the foot of Honister Crag (Britten and Holland); 570 yards (Watson).

759. *Veronica hederifolia*, L. Native. British type. Range 1. Cultivated ground. Rare; ascending to Hallsteads, over Ullswater. By the river Caldew near Rose Castle;

also on hedge-banks at Hensingham near Whitehaven.—(W. Hodgson.)

760. *Veronica agrestis*, L. (Field Speedwell). Native. British type. Range 1. Cultivated ground. Frequent; ascending to 300 yards at Shap.

761. *Veronica polita*, Fries. Native. British type. Range 1. Cultivated ground. Not infrequent; ascending from wall-tops on the esplanade at Grange-over-Sands to 300 yards at Shap.

762. *Veronica Buxbaumii*, Ten. Colonist. British type. Range 1. Cultivated ground. Now quite frequent; ascending from shore-level at Arnside and Flookborough to 300 yards at Shap and near Penrith Beacon.

763. *Bartsia alpina*, L. Native. Highland type. Range 2.
W. In a field by the side of the road leading from Orton to Crosby Ravensworth, on the right-hand side about half-way up the Scar. Recorded both by Ray and Hudson, and still found there.

765. *Bartsia Odontites*, Huds. (Red Rattle). Native. British type. Range 1. Roadsides and cultivated fields. Frequent; ascending to 300 yards near Shap.

Var. *verna* in a corn-field at Seascale.—(Rev. A. Ley.)

766. *Euphrasia officinalis*, L. (Eye-bright). Native. British type. Range 1-3. Pastures and heaths. Common; ascending to the limestone pavement of Whitbarrow and Hutton-Roof, to 500 yards on Great Gable, and 850 yards on Helvellyn.

Var. *gracilis* on Brantfell, Bowness, and in Furness near Goatswater and Cartmel.—(Miss Hodgson.) A large-flowered variety occurs near the 'White Water Dash,' north-west of Skiddaw.—(W. Hodgson.)

767. *Rhinanthus Crista-galli*, L. (Yellow Rattle; Henpens). Native. British type. Range 1-2. Meadows and pastures. Common; ascending to 600 yards.—(Watson.)

Var. *major* at Chapel Bank, St. Helens.—(W. Dickinson.)

770. *Melampyrum pratense*, L. (Cow Wheat). Native. British type. Range 1-3. Woods and grassy heaths. Frequent; ascending from near shore-level between Grange and Lindale to 650 yards on Grisedale Pike.—(Watson.)

Var. *montanum*, Johnst., on a hill south of Ennerdale Lake. —(Rev. F. Addison.)

771. *Melampyrum sylvaticum*, L. Native. Scottish type. Range 1. Woods. Rare.

C. Wilton and Haile near Egremont.—(J. Robson.) Vale of Lorton at Scale Hill.—(J. Woods.) Upland woods over Ullswater.—(W. Hodgson.)

W. Rydal.—(J. Woods.) Barrowfield Wood, Kendal, and on the east side of Whitbarrow.—(T. Gough.)

L. Woods at Yewdale Beck, Coniston.—(Miss Beever.)

773. *Pedicularis palustris*, L. (Lousewort). Native. British type. Range 1-2. Peaty swamps. Frequent; ascending from Newton Regny Moss near Penrith to 450 yards.—(Watson.)

773. *Pedicularis sylvatica*, L. (Small Lousewort). Native. British type. Range 1-2. Damp grassy places. Frequent; ascending to 500 yards on Stake Pass; 540 yards.—(Watson.) A white variety near Shap Abbey.

774. *Scrophularia nodosa*, L. (Stinking Roger; Knotted Figwort). Native. British type. Range 1. Woods and stream-sides. Frequent; ascending in Borrowdale to 250 yards on Castle Crag, and as high in Troutbeck Valley; to 300 yards.—(Watson.)

775. *Scrophularia Balbisii*, Horn. (Marsh Figwort). Native. English type. Range 1. Not seen in the heart of the Lake country, but occurs at Carlisle, Kirkby Lonsdale, and Furness, as at Alithwaite, Bardsea, and below Humphrey Head.

778. *Digitalis purpurea*, L. (Foxglove). Native. British type. Range 1-2. Woods and hillsides. Everywhere common; marking with *Pteris* the upper boundary of the Agrarian region. Mr. Watson notes it at 570 yards. I have seen it on Styhead Pass at 550 yards. Miss Hodgson notes a white variety in Furness at Haverthwaite, and a monstrosity with polypetalous corolla at Rosshead fields.

Antirrhinum majus, L. (Snapdragon). Alien. Sometimes naturalised on old walls, as at Townend, at the foot of the Winster valley.

780. *Antirrhinum Orontium*, L. Colonist. English type. Range 1.

C. Hedge-banks at Gosforth.—(J. Robson.) Plentiful about Braystones, whence it was sent to the Exchange Club in 1865 by Mrs. Pratten, the daughter of Knapp, who wrote on grasses, and after whom the genus *Knappia* is named. Mrs. Pratten's plants are now in the Museum at Whitehaven. Scalelands.—(Whitehaven Cat.)

Linaria Cymbalaria, Mill. Alien. Frequently established on old walls, as at Saurey, Grange-over-Sands, Rydal, Calder Abbey, Rose Castle, and the bridge at Dallam Tower near Milnthorpe.

784. *Linaria repens*, Ait. Native. English type. Range 1.
C. Hedgerow at Buckabank; introduced.—(W. Hodgson.)
W. Winster Valley, Windermere.—(F. Clowes.) At a cave called the Cow's Mouth, Morecambe shore near Arnside Knot. —(J. H. Rossall.)

L. Rocks by the roadside at Nibthwaite, Coniston Water, a variety with white unstriped flowers; seen also at Newby Bridge, and in gardens at Ambleside.—(Borrer, Phytol. ii. 426.) This has been called *Linaria italica.*

785. *Linaria vulgaris*, Mill. (Toadflax). Native. British type. Range 1. Dry banks. Frequent; ascending from the shore at Flookborough and Humphrey Head to 200 yards on Latrigg and the foot of Brantfell, Bowness, and Ullswater shore, Gowbarrow.

787. *Linaria minor*, Desf. Colonist. English type. Range 1.

C. Plentiful on the railway near Brigham station; also near Aspatria, among the ballast on the line; spreading rapidly at both stations.—(W. Hodgson.)

W. Railway bank between Staveley and Ings, Kendal.— (T. Gough, J. H. Lewis.)

Sibthorpia europæa, L. This is given as a plant of the Westmoreland hills in several of the old books, no doubt by some blunder, perhaps by confusion with *Hydrocotyle.*

Mimulus luteus, L. (Monkey Plant). Alien. Cultivated everywhere in gardens up to the Kirkstone Inn, and now thoroughly established in swamps and about streams in many places, especially in Fusedale Beck near Howtown. Also at Seascale, Matterdale, Stainton, Bardsea, Newby Bridge, and Kirkby Lonsdale.

ORDER OROBANCHACEÆ.

797. *Lathræa squamaria*, L. (Toothwort). Native. English type. Range 1. Woods, especially of hazel. Not infrequent.

C. Akebank near Wigton.—(Rev. J. Dodd.) Woodhall,

Cockermouth.—(G. Mawson.) Near Lyulph's Tower, Ullswater.—(W. Hodgson.) Banks of the Lowther at Askham.—(W. Hodgson.)

W. Casterton Woods, Kirkby Lonsdale. — (Hindson.) Winder Scar, Scout Scar, and Cunswick Wood, near Kendal. First recorded by Wilson. Eamont banks between Yanwath and Eamont Bridge.—(Mrs. King.) Pooley Bridge.—(J. B. Davies.) Ascending Wansfell from Ambleside.—(F. Clowes.) Between Rydal and Ambleside.—(F. Jones.) Drawn from the Nook, Ambleside, by Miss Wilson.

L. Lake-side at Coniston.—(Linton.)

I have no record of any *Orobanche* from the district except a note from Mr. W. B. Waterfall that *O. major* has been reported from Ullock near Whitehaven.

ORDER VERBENACEÆ.

798. *Verbena officinalis*, L. (Vervain). Denizen. English type. Range 1. Roadsides near farm and villages. Rare.

C. Irton.—(J. Robson.) Plentiful at Cockermouth.—(T. Lawson.)

W. Abundant on Whitbarrow.—(F. Clowes.) Roadside at Clawthorpe.—(B.)

L. In Furness at Lindale, Grange, and Alithwaite.—(B.) Near Newland, Ulverstone.—(Miss Hodgson.)

ORDER LABIATÆ.

801. *Lycopus europæus*, L. (Gipsy Wort). Native. British type. Range 1. Marshes and stream-sides. Rare.

C. Ribton Hall, Workington.—(Mr. Tweddle). Sellafield Tarn near Gosforth.—(J. Robson.) Drigg Moor, Ravenglass.—(Linton.) In the pinfold at Aspatria; also in Salta Moss near Dubmill.—(W. Hodgson.)

W. Round Windermere in a few places.—(F. Clowes.) Drawn from Skelwith by Miss Wilson. Marsh below Middlebarrow Wood, Arnside.—(B.) Burnside Farm, Kendal.— (Linton.)

L. In the moss at the head of Esthwaite Water.— (T. Lawson.) Hollows of the coast sand-hills at Ronnard.— (Dr. F. A. Lees.)

802. *Mentha rotundifolia*, L. Native. English type. Range 1.

C. In Borrowdale at Lodore, below the waterfall, and near the Bowder Stone. First recorded by Winch. Naddle Beck, Keswick.—(W. Dickinson.) In Patterdale village, near the inn.—(W. Foggitt.) Portinscale, Keswick.—(Rev. A. Ley.) Abundant on the north shore of Ullswater, from Floshgate to Pooley Bridge; first recorded by Professor Balfour.

L. Near Grange-over-Sands.—(T. Lawson.)

Mr. J. C. Melvill believes that he saw *M. sylvestris* at Furness Abbey in 1865, but kept no specimen.

Mentha viridis, L. Alien.

C. Rosley near Wigton.—(Rev. R. Wood.)

W. Seen by Mr. Watson at Crosby Ravensworth; not native.

805. *Mentha piperita*, Huds. (Pepper Mint). Native. English type. Range 1.

C. Roadside below the Cove farm, Watermillock, and in a branch of the Eamont near Pooley Bridge.—(W. Hodgson.) By the Ehen at Egremont.—(Whitehaven Cat.) Naddle near Keswick.—(W. Dickinson.)

W. Whitbarrow.—(F. Clowes.) Roadside in Little Langdale.—(J. S. Mill!)

806. *Mentha hirsuta*, L. (Common Mint). Native.

British type. Range 1. Swamps and stream-sides. Common; ascending to 450 yards.

Var. *subglabra*, in Borrowdale by a stream half a mile south of Grange (C. Bailey); and in the marsh below Middlebarrow Wood, Arnside (B.).

807. *Mentha sativa*, L. Native. British type. Range 1. Swamps and ditches. Common; up to 250 yards in Troutbeck Valley.

Var. *paludosa*, as common as the type. Var. *rubra*, in Furness, by the river Crake.—(Miss Hodgson.) Var. *gentilis*, in the Naddle valley near Keswick.—(W. Dickinson.)

808. *Mentha arvensis*, L. (Field Mint). Native. British type. Range 1. Cultivated fields. Frequent; up to 300 yards over Penrith.

809. *Mentha Pulegium*, L. (Penny Royal). Native. English type. Range 1.

C. Saltcoats.—(Rev. R. Wood.)
L. At Goose Green, near Dalton in Furness.—(Aiton.)

810. *Thymus Serpyllum*, L. (Thyme). Native. British type. Range 1-4. Rocks and dry banks. Frequent at all elevations, both amongst the limestone and slate hills; ascending to the summit of Whitbarrow and Hutton-Roof, to 850 yards on Helvellyn, and to 1000 yards on Scawfell Pike.

811. *Origanum vulgare*, L. (Wild Marjoram). Native. Xerophilous. British type. Range 1. Woods and dry banks; almost confined to the limestone.

C. Slapestones brow, Penrith, and by the road leading thence to Stainton.—(W. Hodgson.)
W. Woods about Lowther, Clibburn, and Great Strickland.

—(B.) Abundant near Arnside Tower.—(B.) Cunswick Wood, Kendal.—(T. Gough.)

L. Islands of Windermere.—(W. Foggitt.) Woods at Grange-over-Sands.—(B.) In a wood near Cartmel Wells.—(Mr. Atkinson.) Bardsea Park.—(Aiton.)

812. *Calamintha Acinos*, Clairv. (Basil). Native. British type. Range 1. Dry banks. Rare.

C. Lower Lingbank, Nethertown.—(M. Chambers.) Recorded by mistake as *Stachys annua* in Martineau's Guide. Fields at Ainstable.—(Winch.) Roadside hedge-bank between New Cooper and Hangingshaw Moss, scarce.—(W. Hodgson.)

W. Sandy field at Clibburn.—(B.) Glenridding Valley.—(W. Foggitt.) Scout Scar, Kendal.—(T. Gough.)

L. In plenty on the top of Humphrey Head.—(C. J. Ashfield.) Railway bank west of Cark station.—(B.) Railway bank at Foxfield Junction.—(Rev. A. Ley.)

814. *Calamintha officinalis*, Moench. (Calamint). Native. English type. Range 1. Dry banks. Very rare.

C. Near Calva Hall.—(W. Dickinson.)

W. On the walls of Kendal Castle.—(J. Wilson, T. Gough.)

L. Railway embankment at Grange-over-Sands.—(B.)

815. *Calamintha Clinopodium*, Spenn. (Wild Basil). Native. British type. Range 1. Woods and thickets. Frequent; ascending from shore-level at Flookborough and Humphrey Head to Lodore and Lowther Woods; 300 yards.—(Watson.)

818. *Teucrium Scorodonia*, L. (Wood Sage). Native. British type. Range 1-2. Woods and rocky hillsides. Frequent; ascending to the top of Hutton-Roof and Whitbarrow, and to 400 yards in Great Langdale.

822. *Ajuga reptans*, L. (Bugle). Native. British type. Range 1-2. Damp grassy places. Frequent; ascending to 560 yards.—(Watson.)

823. *Ajuga pyramidalis*, L. Native. Scottish type. Range 2.

W. Very fine on precipitous rocks of Hill Bell.—(J. Backhouse.)

825. *Ballota nigra*, L. (Black Horehound). Native. English type. Range 1. Only seen in waste ground at Workington.—(Mr. Tweddle.) Workington marsh side.—(W. Dickinson.)

Leonurus cardiaca, L. (Motherwort). Alien.

C. Waste ground at Workington Row.—(Mr. Tweddle.) A drawing in Rooke's Flora, 'Lane behind Dundraw, 1851.'—(W. Hodgson.) Near Curthwaite.—(Rev. R. Wood.)

W. In a farmyard at Whitbarrow.—(Rev. A. Bloxam.)

827. *Lamium Galeobdolon*, Crantz. (Archangel). Native. English type. Range 1. Shaded woods. Very rare.

C. Crosedale and Ennerdale.—(J. Robson.) Portinscale, Keswick.—(Whitehaven Cat.) Wood between Penrith and Edenhall.—(Mrs. F. King.)

L. Reported from Coniston, but not seen by Miss Beever.

828. *Lamium album*, L. (White Dead Nettle). Native. British type. Range 1. Not seen in the interior of the Lake country, about Keswick, Ambleside, or Coniston, but frequent at Whitehaven, Cartmel, Arnside, and Kirkby Lonsdale. A casual at Watermillock over Ullswater, 300 yards.—(W. Hodgson).

Lamium maculatum, L. Alien. An occasional straggler from gardens, as at Ullock near Whitehaven, Sawrey, Kirkby

Lonsdale, between Witherslack Hall and Townend, and near St. Paul's Chapel in Winster Valley.

830. *Lamium amplexicaule*, L. Native. British type. Range 1.

C. Gardens and rubbish-heaps about Ullswater, scarce.—(W. Hodgson.) Brigham near Cockermouth.—(W. B. Waterfall.)

831. *Lamium purpureum*, L. (Red Dead Nettle). Native. British type. Range 1. Hedge-banks and cultivated fields. Frequent; ascending to 250 yards near Windermere railway station, and 300 yards at Shap and Penrith.

831*. *Lamium incisum*, Willd. Native. British type. Range 1. In similar places to the last, but much less common. Brackenthwaite, Westward, Ullswater, Newton Regny, Kendal, Holme, Arnside, Bardsea, Ulverstone, Grange-over-Sands, etc.

832. *Galeopsis Ladanum*, L. Colonist. English type. Range 1. Cultivated fields. Rare.

C. St. Bees.—(J. Robson.)

W. At Lansmoor near Great Strickland.—(Lawson.) Foot of Scout Scar, Kendal.—(T. Gough.) Arnside Knot.—(Prof. Oliver.)

L. Fields at Hawkshead.—(Linton.)

834. *Galeopsis Tetrahit*, L. (Hemp Nettle). Native. British type. Range 1-2. Cultivated fields; one of the commonest weeds; ascending to 250 yards in Troutbeck Valley, in Borrowdale to Stonethwaite, 300 yards over Coniston; 350 yards.—(Watson.)

835. *Galeopsis versicolor*, Curt. (*G. speciosa*, Miller). Colonist. Scottish type. Range 1. Cultivated fields. Rare.

C. St. Bees.—(J. Robson.)

W. In fallow ground near Hutton-Roof.—(Mr. Atkinson.) Sprint Bridge and Burnside Hall near Kendal.—(T. Gough.)

L. Hedges at Kirby in Furness.—(Atkinson.) In Leighton Park, close to a deep drain across the bog.—(Miss Beever.)

836. *Stachys Betonica*, Benth. (Betony). Native. English type. Range 1-2. Frequent in pastures, ascending to 250 yards in the Watendlath and Troutbeck valleys; 360 yards in Mardale.—(Watson.)

837. *Stachys palustris*, L. (Marsh Woundwort). Native. British type. Range 1. Stream-sides and roadsides. Frequent, from the shore at Flookborough up to 250 yards in Troutbeck Valley and over Penrith.

837*b*. *Stachys ambigua*, Sm. Native. British type. Range 1. A hybrid. Frequent where *S. palustris* and *sylvatica* grow together.

C. In Borrowdale at Seatollar, and fine at Seathwaite.—(B.) Roadside near Buttermere village.—(B.) Pardshaw.—(W. B. Waterfall.)

W. In the grounds of the Ullswater Inn at the foot of Glenridding.—(B.) Rubbish-heaps at the south end of Holme village.—(B.) Roadsides at Bowness, Clappersgate, and near Windermere Water-head.—(B.) Roadside at Thrimby near Shap.—(B.)

L. Two places by the road below Sawrey, towards the Ferry Inn.—(B.) Garden fence at Dalton in Furness.—(Dr. F. A. Lees.) About the lower slate quarries in Coniston village.—(B.)

838. *Stachys sylvatica*, L. (Wood Woundwort). Native. British type. Range 1. Woods and thickets. Common;

ascending to 250 yards in Borrowdale on Castle Crag ; 300 yards.—(Watson.)

840. *Stachys arvensis*, L. Colonist. British type. Range 1. Cultivated fields. Frequent, up to Grange in Borrowdale, Ennerdale Lake, and above Lowthwaite in the Vale of St. John.

841. *Nepeta Glechoma*, Benth. (Ground Ivy). Native. British type. Range 1. Frequent on hedge-banks ; ascending from shore-level at Flookborough to 300 yards.—(Watson.)

842. *Nepeta Cataria*, L. (Cat Mint). Native. English type. Range 1.

C. Waste ground near the mill at Dalemain.—(W. Hodgson.) By the road from Bell Bridge to Sebergham Hall.—(W. Duckworth.)

L. On the beach at Rampside in Furness.—(Mr. Atkinson.)

Marrubium vulgare, L. (Horehound). Alien. Waste ground. Very rare.

C. Baggrow.—(Whitehaven Cat.)

L. Near Jacklands Tarn, Low Furness.—(Miss Hodgson.)

844. *Prunella vulgaris*, L. (Self-heal). Native. British type. Range 1-3. Grassy places. Common ; ascending to 500 yards on Great Gable, 600 yards on High Street, and in Tongue Ghyll, between Seat Sandal and Fairfield ; 680 yards.—(Watson.)

845. *Scutellaria galericulata*, L. (Skull-cap). Native. British type. Range 1. Lake- and stream-sides. Not uncommon. Lowes Water, Windermere, Grasmere, Coniston Lake, by the Eamont at Pooley Bridge, banks of the Ehen, Dub Beck near Cleator, Braithwaite Beck near Keswick, etc.

846. *Scutellaria minor*, L. Native. English type. Range 1.

C. Black Moss, Wormgill, Drigg, and Ennerdale.—(Whitehaven Cat.) Dent Hill.—(Rev. F. Addison, W. Hodgson.) Ghyll, Egremont.—(J. Robson.) Moor near Whitehaven.—(Mrs. Blackett.) Margins of Wastwater and Crummock Lake.—(Black's Guide.) Ladstocks in Thornthwaite.—(W. Dickinson.)

W. Windermere, in some of the bogs.—(F. Clowes.)

L. Hawkshead Mill, and about the Coniston tarns, and by the stream below Tarn Hause.—(Miss Beever.)

ORDER BORAGINACEÆ.

847. *Myosotis palustris*, With. (Forget-me-not). Native. British type. Range 1. Ditches and stream- and lake-sides in the low country. Frequent; ascending to 250 yards at Rossgill and near Thrimby Hall. Very fine about the Brathay at Skelwith.

Var. *strigulosa*, Reich., found near Keswick by Mr. J. B. Davies.

848. *Myosotis repens*, D. Don. Native. British type. Range 1-3. Swamps and ditches. Frequent; ascending to 800 yards at Red Tarn, and 600 yards in the sykes over Grisedale Tarn. Very fine below Watermillock, where Mr. Hodgson showed it to me in September 1883.

849. *Myosotis cæspitosa*, Schultz. Native. British type. Range 1-2. Swamps and ditches. Frequent; ascending from the shore marshes at Arnside and Flookborough to 500 yards over Hayes Water.

851. *Myosotis sylvatica*, Ehrh. Native. English type. Range 1. Dense woods. Rare.

C. St. Herbert's Island, Derwentwater.—(Winch.) Aspatria

near Allonby.—(W. B. Waterfall.) A drawing in Rooke's Flora, marked 'Stanley Burn, 1850.'

W. Buckbarrow Scar, Long Sleddale.—(Wilson.)

852. *Myosotis arvensis*, Hoffm. Native. British type. Range 1. Frequent in cultivated fields; ascending to 300 yards amongst the limestone cliffs of Shap Common and on Castle Crag, Borrowdale.

Var. *umbrosa* in woods near the Ferry Inn, Windermere.

853. *Myosotis collina*, Hoffm. Native. British type. Range 1. Dry banks. Not infrequent in the lower zone; ascending to the churchyard at Watermillock, 300 yards.—(W. Hodgson!)

854. *Myosotis versicolor*, Lehm. Native. British type. Range 1. In similar places to the last. Not uncommon. Whitehaven, Ullswater, Walney Island, Colton in Furness, Newby Bridge, Bowness, Kirkby Lonsdale, etc. Drawn from Loughrigg Tarn by Miss Wilson.

855. *Lithospermum officinale*, L. (Gromwell). Native. Xerophilous. British type. Range 1. Woods in the limestone districts.

C. At Mosser, and near Lorton and Westward Park.—(W. Dickinson.)

W. Barrowfield Wood, Kendal.—(T. Gough.) Roadside below St. Paul's Church, Winster Valley.—(B.) Winster Valley, east side.—(W. Matthews.) Wood near Arnside station.—(W. Foggitt.) About Wrayton and Thurland Castle.—(Hindson.)

L. Furness Abbey.—(W. Foggitt.) Woods between Grange and Lindale.—(B.) About Plumpton, and on the beach at Bardsea.—(Miss Hodgson.)

856. *Lithospermum arvense*, L. Colonist. British type. Range 1. Cultivated fields. Rare.

C. Stanger.—(W. Dickinson.)

W. Corn-fields near Kendal.—(T. Gough.) On Lansmoor near Great Strickland.—(T. Lawson.)

858. *Mertensia maritima*, G. Don. Native. Maritime. Scottish type. Range 1.

C. Here and there between Maryport and Workington.—(Rev. J. Harriman.) Parton shore.—(Glaister and Leitch.) Near Whitehaven.—(T. Lawson.) Between Ravenglass and Bootle.—(Wood, J. Robson.)

L. Plentiful on Walney Island, over against Biggar.—(T. Lawson.) Seen there in 1881.—(Rev. A. Ley.)

859. *Symphytum officinale*, L. (Comfrey). Denizen. English type. Range 1.

C. Banks of Lowea Beck.—(Whitehaven Cat.)

W. Not uncommon about Windermere.—(F. Clowes.) Drawn from Troutbeck by Miss Wilson. By the side of the road to Kendal near Bowness.—(F. C. Roper.) Roadside near Storrs Hall, Windermere.—(B.) Banks of the Lune near Kirkby Lonsdale.—(B. D. Jackson.)

860. *Symphytum tuberosum*, L. Denizen. Intermediate type. Range 1.

C. By the Furness railway, Green Bank, Whitehaven.—(Whitehaven Cat.)

W. Wood at the foot of Loughrigg Fell, on the banks of the Rothay, between Ambleside and Rydal.—(C. Bailey.) Drawn from Fox How by Miss Wilson.

Borago officinalis, L. (Borage). Alien. An occasional straggler from gardens. St. Bees, Patterdale village, Aldingham, Kirkby Lonsdale, and at the foot of Ullswater near Floshgate.

Borago orientalis, L. Alien.

C. Wood by the roadside at Portinscale near Keswick.—(Borrer.)

862. *Lycopsis arvensis*, L. (Bugloss). Colonist. British type. Range 1. Cultivated fields in sandy ground. Rare.

C. St. Bees.—(W. Dickinson.) Maryport.—(Whitehaven Cat.)

L. Rampside in Furness.—(Miss Beever.)

Anchusa sempervirens, L. (Alkanet). Alien. An occasional straggler from gardens. Whitehaven, Swarthmore Hall, Gosforth, Sandwith, St. Bees, near Loughrigg Tarn, Patterdale village, Pooley Bridge, Lowther, Kendal, etc.

866. *Cynoglossum officinale*, L. (Hound's Tongue). Native. English type. Range 1. Woods and thickets. Rare.

C. Hawksdale Bridge, Carlisle.—(W. Hodgson.) Dalston.—(W. Duckworth.)

W. At Sizergh and Levens near Milnthorpe.—(T. Gough.)

L. Furness shores.—(Miss Hodgson.) Coast sand-hills of the Duddon estuary at Ronnard.—(Dr. Lees.)

Pulmonaria officinalis, Lungwort, *angustifolia*, *et Omphalodes verna*, are all three occasional escapes from cottage gardens.

869. *Echium vulgare*, L. (Viper's Bugloss). Native. British type. Range 1. Dry banks. Rare.

C. Aspatria and Whitehaven bleach-green.—(Whitehaven Cat.) Railway bank at Coulderton.—(J. Robson.) Scotby near Carlisle.—(D. Oliver.) St. Bees.—(W. B. Waterfall.)

W. Corn-fields at Kendal.—(T. Gough.)

ORDER PINGUICULACEÆ.

872. *Pinguicula vulgaris*, L. (Butterwort). Native. Scottish type. Range 1-3. Moory swamps. Frequent; ascending to 700 yards on Helvellyn and Scawfell Pike, 600 yards on High Street. A variety *longicornis* gathered by Woods in Fisher Place Ghyll, between Great Dod and Helvellyn.—(Phytol. i. 310.)

875. *Utricularia vulgaris*, L. (Bladderwort). Native. British type. Range 1. Swamps and ponds. Not infrequent. Maryport, Windermere, Derwentwater, Brigstear Moss near Kendal, Newton Regny Moss near Penrith, Terra Bank Tarn near Kirkby Lonsdale, Keswick Tarn near Ulverstone, etc.

876. *Utricularia intermedia*, Hayne. Native. Local type. Range 1. Moory ponds. Rare.

C. Ditch at the foot of Derwentwater.—(Winch.) Boggy ground at head of Ennerdale Lake.—(J. Adair, Rev. A. Ley.)

L. Between Brathay and Hawkshead.—(Rev. F. J. A. Hort.) In a pool near Coniston High Cross, off the road on the left-hand side going from Coniston.—(Miss Beever.)

877. *Utricularia minor*, L. (Little Bladderwort). Native. British type. Range 1. Moory pools and ditches.

C. Ditch near Dubmill, Allonby.—(W. Hodgson.) Ditches on the west side of Derwentwater.—(Black's Guide.) Stonethwaite Moss, Thirlmere, and Eskmeals, Ravenglass.—(W. Dickinson.) Wedholme Flow, Wigton.—(W. B. Waterfall.) At Cooper, in Bromfield parish.—(Rev. J. Dodd.) Ullock Moss, Keswick.—(Miss Edmonds.) Boggy ground, Dent Hill.—(Rev. F. Addison.) Bog-holes in Ennerdale.—(Britten and Holland.) Eel Tarn near Wastwater.—(Rev. A. Ley.)

W. Brigstear Moss, Kendal.—(T. Lawson.) Swindale near Shap.—(Watson.) Clibburn Moss, and pools on Barton Fell, Ullswater.—(W. Hodgson.)

L. Ditches at Outerthwaite near Flookborough.—(Rev. Mr. Jackson.) In a tarn between Hawkshead and Coniston, with the last species.

ORDER PRIMULACEÆ.

878. *Primula vulgaris*, Huds. (Primrose). Native. British type. Range 1-2. Woods and thickets. Common everywhere in the lower zone; ascending to the limestone pavement of Whitbarrow, and 400 yards in Great Langdale.

880. *Primula veris*, L. (Cowslip). Native. British type. Range 1. Common in pastures throughout the lower zone. The hybrid oxlip (*P. variabilis*, Goupil, Lady Candlestick; Cow Sinkin, often in times past mistaken for *P. elatior*, Jacq.) is common throughout the Lake district.

881. *Primula farinosa*, L. (Mealy Primrose). Native. Intermediate type. Range 1-2. Swampy fields. Common about Windermere, and round Ullswater, ascending up Troutbeck to the top of Kirkstone Pass (500 yards). Plentiful also about Kendal, Shap, and Kirkby Lonsdale. It occurs also on Arnside Knot, Catlands near Wigton, in St. John's Vale, and in a few places in West Cumberland, but is not found on the Keswick, Penrith, and Coniston hills, and Miss Hodgson altogether omits it from her catalogue of Furness plants.

Trientalis europæa, L. Hudson specially mentions this as a Westmoreland plant, and Aiton says it has been found amongst the hills in the north-west of Furness. It is frequent in North Yorkshire.

885. *Hottonia palustris*, L. (Water Violet). Native. English type. Range 1.

W. In the river Kent and the ditches of Brigstear Moss near Kendal. First recorded by Lawson. Now extinct.

L. In the mill-pond at Bardsea.—(Miss Hodgson.)

886. *Lysimachia vulgaris*, L. (Yellow Loosestrife). Native. English type. Range 1. Lake-sides and stream-sides. Not infrequent. Lowes Water, Windermere, Ullswater, Coniston Water, Derwentwater, Lorton, Ennerdale, Urswick Tarn, Penrith, Greystoke, banks of the Ehen, etc. Drawn from Pull Wyke by Miss Wilson.

Lysimachia ciliata, L. Alien.

C. Roadside at Sebergham, where the late Rev. R. Wood said he had known it more than sixty years, and reported also by Mr. Wright from a slate quarry on Warnell Fells. Still at the Sebergham station, and likely to be permanent, 1883.—(W. Hodgson.)

887. *Lysimachia thyrsiflora*, L. Native. Intermediate type. Range 1.

C. Sellafield Tarn near Gosforth.—(J. Robson.)

888. *Lysimachia nummularia*, L. (Moneywort). Native. English type. Range 1.

C. Ditch on the east slope of Latrigg.—(B.) Irton Wood, Wastdale.—(J. Robson.) Bolton branch railway at Baggrow.—(Whitehaven Cat.)

W. Shores of Windermere at Bowness, etc.—(F. Clowes.) Not uncommon at Kirkby Lonsdale.—(Hindson.) Drawn from Fox Ghyll by Miss Wilson.

L. Bardsea mill-pond and Pull Wyke, Windermere.—(Miss Hodgson.)

ORDER PRIMULACEÆ. 173

889. *Lysimachia nemorum*, L. Native. British type. Range 1-2. Swamps and damp woods. Common; ascending to 400 yards in Great Langdale, 500 yards at Hayes Water; 560 yards.—(Watson.)

890. *Anagallis arvensis*, L. (Field Pimpernel). Colonist. British type. Range 1. Cultivated ground and wall-tops. Frequent.

Var. *cærulea* at Hensingham toll-bar near Whitehaven (W. Dickinson); and Roosebeck in Furness (R. Ashburner); and a form with reddish-brown flowers at Colton in Furness (Miss Hodgson). There is a drawing of *A. cærulea* in Rooke's Flora, marked 'Workington Hall Park.'

891. *Anagallis tenella*, L. (Bog Pimpernel). Native. British type. Range 1-2. Swampy hillsides. Frequent; ascending in Little Langdale to Blea Tarn, and 350 yards on Coniston Old Man.

892. *Centunculus minimus*, L. Native. English type. Range 1.

C. On the coast at Ravenglass.—(J. Woods.)

L. Salt marshes and meadows by the seaside at Newton in Cartmel.—(Mr. Hall.) Modern confirmation wanted.

893. *Samolus Valerandi*, L. (Brook-weed). Native. English type. Range 1.

C. Shore at Coulderton, St. Bees, and Waberthwaite.—(W. Dickinson.) Fleswick Beck, south of Whitehaven.—(Whitehaven Cat.)

W. Formerly plentiful on Brigstear Moss near Kendal; now extinct.—(T. Gough.) Marsh below Middlebarrow Wood, Arnside.—(B.)

L. On the Furness shore at Plumpton, Flookborough, Kents

Bank, and Grange. First recorded by Lawson. Damp sandy hollows of the coast at Ronnard.—(Dr. F. A. Lees.)

894. *Glaux maritima*, L. (Saltwort). Native. Maritime. British type. Range 1. Salt marshes along the coast. Common. St. Bees, Ravenglass, Seascale, Walney Island, Ulverstone, Flookborough, Grange, Arnside, etc.

ORDER PLUMBAGINACEÆ.

895. *Armeria maritima*, Willd. (Thrift). Native. British type. Range 1-4. Common all along the coast from St. Bees to Arnside. Inland at 800-900 feet on Scawfell, where it is called the Scawfell Pink. Pierce Ghyll.—(Melvill.) Cliffs of the east face of Helvellyn. On Dove Crags, Grassmoor (W. B. Waterfall), and hills between the top of the Vale of Newlands and Borrowdale (Watson).

897. *Statice Limonium*, L. (Sea Lavender). Native. Maritime. English type. Range 1. Salt marshes at St. Bees, Barrow, and Grange-over-Sands. Not plentiful.

898. *Statice bahusiensis*, Fries. Native. Maritime. English type. Range 1.

L. Slate rocks on the shore at Greenodd near Ulverstone. —(Miss Hodgson.) Salt marsh, Isle of Walney, with *Œnanthe Lachenalii*.—(Dr. F. A. Lees.)

899. *Statice binervosa*, G. E. Sm. Native. Maritime. Atlantic type. Range 1.

C. Rocks of St. Bees Head.—(W. Dickinson, etc.) First wrongly referred to *reticulata* in the Botanist's Guide. Fleswick (misprinted Keswick).—(Whitehaven Cat.)

L. Salt marsh between Tridley and Grenodd, abundant. —(Miss Hodgson.)

ORDER PLANTAGINACEÆ.

901. *Plantago major*, L. (Way Bread; Common Plantain; Rat-tails). Native. British type. Range 1-2. Roadsides. Common; ascending to 400 yards on Coniston Old Man, 500 yards on Kirkstone Pass; 520 yards.—(Watson.)

902. *Plantago media*, L. (Lamb's Tongue). Native. English type. Range 1. Dry banks. Frequent, especially on the limestone, ascending to the top of Whitbarrow and Hutton-Roof, and 360 yards near Shap.—(Watson.)

903. *Plantago lanceolata*, L. (Ribwort). Native. British type. Range 1-3. Grassy places. Common; ascending to 500 yards at Hayes Water and on Kirkstone Pass; 650 yards. —(Watson.)

Var. *altissima* seen at Grange-over-Sands.

904. *Plantago maritima*, L. (Sea Plantain). Native. British type. Range 1-2. Common all along the coast. Inland near Gillerthwaite in Ennerdale (J. Robson); and at the head of Fusedale (W. Hodgson).

905. *Plantago Coronopus*, L. (Buck's Horn Plantain). Native. Maritime. British type. Range 1. Along the shore. Frequent. St. Bees, Flimby, Ravenglass, Seascale, Walney Island, Cark, Milnthorpe estuary, etc.

906. *Littorella lacustris*, L. (Shore-weed). Native. British type. Range 1. Lakes and tarns. Frequent, up to 500 yards at Hayes Water, Styhead Tarn, and Blea Tarn, Watendlath.—(W. Foggitt.)

ORDER CHENOPODIACEÆ.

Chenopodium olidum, Curt. Alien.

C. On the Solway shore at Saltcoats, beyond our bounds. —(Rev. R. Wood.)

L. Waste ground at Barrow in Furness.—(W. Foggitt.) Mrs. F. King showed me *C. polyspermum* growing as a weed in her garden at Penrith; introduced from the south of England. A plant in my possession came up in a garden at Thorncroft, Workington.—(W. Hodgson.)

911. *Chenopodium rubrum*, L. Native. English type. Range 1.

W. Frequent on dunghills about Kirkby Lonsdale.— (Hindson.)

914. *Chenopodium album*, L. (Meals; White Goose-foot). Native. British type. Range 1. Common in cultivated ground, ascending to 300 yards over Penrith and Bowness.

917. *Chenopodium Bonus-Henricus*, L. (Good King Henry). Denizen. British type. Range 1. Common about villages and farm-houses; ascending to 300 yards at Shap, and on Gowbarrow Fells; very common about Hawes Water, Shap, and Rossgill. Sold in Penrith market as 'Mercury.'—(Britten and Holland.)

918. *Obione portulacoides*, Moq. Native. Maritime. English type. Range 1.

L. In a brackish ditch on Walney Island, opposite Barrow. —(C. Bailey.) On the coast at Barrow.—(D. Oliver.) Saltmarsh at Kents Bank, and on the railway embankment west of Cark station.—(B.)

ORDER CHENOPODIACEÆ.   177

920. *Atriplex arenaria*, Woods.   Native.   Maritime. British type.   Range 1.

C. On the shore at Flimby near Maryport.—(W. Hodgson!) Allonby.—(Rev. R. Wood.)   Drawn from St. Bees in Rooke's collection.

921. *Atriplex Babingtonii*, Woods.   Native.   Maritime. British type.   Range 1.

C. On the seashore at Lowea near Whitehaven.—(Rev. F. Addison.)   Near Dubmill, Allonby.—(W. Hodgson.) Coulderton.—(Rev. R. Wood.)

L. On the shore of the Leven estuary at Roosebeck and Greenodd.—(Miss Hodgson.) On the shore at Grange, west of Humphrey Head, and at Flookborough and Cark.—(B.)

922. *Atriplex hastata*, L. (Orache).   Native.   British type. Range 1-2. Common in cultivated ground; ascending as high as cultivation reaches (500 yards) on Kirkstone Pass. The common Lakeland form is *A. Smithii*, Syme. I have not seen *A. deltoidea* at the Lakes. *A. triangularis*, Willd., is reported from St. Bees in the Whitehaven Catalogue.

923. *Atriplex angustifolia*, L.   Native.   British type. Range 1. Common in cultivated ground; ascending from coast-level in Furness to 300 yards at Shap. *A. erecta*, Huds., is common in rich soil.

924. *Atriplex littoralis*, L.   Native.   Maritime.   British type.   Range 1.

C. Seashore at Parton near Whitehaven.—(Rev. F. Addison.)

925. *Beta maritima*, L. (Wild Beet).   Native.   Maritime. British type.   Range 1.

C. With the last on the seashore at Parton near Whitehaven.—(Rev. F. Addison, W. Hodgson.)

M

926. *Salsola Kali*, L. (Saltwort). Native. Maritime. British type. Range 1.

C. On the shore at Allonby and Coulderton.—(J. Robson, Whitehaven Cat.) On the shore at Lowea near Whitehaven.—(Rev. F. Addison.) The fresh sprouts are sometimes preserved, and known locally as 'Parton Pickle.'—(W. Hodgson.)

L. Roosebeck shores, Furness.—(Miss Ashburner.)

927. *Suæda maritima*, Dum. Native. Maritime. British type. Range 1.

C. On the shore at Coulderton.—(J. Robson.)

L. In Furness, in Plumpton salt-marsh near Ulverstone.—(Miss Ashburner.) Salt-marshes near Cark, and on the shore west of Humphrey Head.—(B.)

929. *Salicornia herbacea*, L. (Glasswort Samphire). Native. Maritime. British type. Range 1.

C. On the shore at Workington and Ravenglass.—(J. Robson.) Cardurnock Point.—(Whitehaven Cat.)

L. On the shore south of Flookborough.—(B.)

Var. *procumbens*, Sm.

C. On the north shore at Workington.—(W. Dickinson.)

L. Salt-marsh of the Duddon estuary at Foxfield, and the Leven estuary at Greenodd.—(Miss Hodgson.) This is wrongly cited as *S. radicans* in Linton's Guide.

ORDER POLYGONACEÆ.

931. *Polygonum Bistorta*, L. (Snakeweed, Bistort). Denizen. British type. Range 1. Frequent throughout the Lake district about villages and farm-houses. Local names, 'Eastermir-giants;' 'Easter-ledge.' The leaves are used as an ingredient in herb puddings.—(W. Hodgson.)

932. *Polygonum viviparum*, L. (Alpine Bistort). Native. Highland type. Range 2-3.

W. Abundant on the hills near Crosby Ravensworth.— (Lawson.) Near the road between Hardendale and Shap.— (J. Wilson.) Cliffs of the east face of Helvellyn, 800-900 yards; first recorded by Balfour.

933. *Polygonum amphibium*, L. Native. British type. Range 1. Frequent in lakes and ponds in the lower zone. Fine examples occur in the old reservoir at Maryport.— (W. Hodgson.)

934. *Polygonum lapathifolium*, L. Native. British type. Range 1. Cultivated fields and rubbish-heaps. Frequent; ascending to 250 yards in Great Langdale.

935. *Polygonum Persicaria*, L. (Redshanks). Native. British type. Range 1-2. Throughout the Lakes, one of the commonest weeds of arable land, up to 300 yards over Penrith; 400 yards.—(Watson.)

936. *Polygonum mite*, Schrank. Native. Germanic type. Range 1.

C. St. Bees Valley.—(J. Robson.) Confirmation wanted.

937. *Polygonum Hydropiper*, L. (Water Pepper). Native. British type. Range 1-2. Ditches and damp ground. Common, ascending to 400 yards in Hag Ghyll, Troutbeck; 450 yards.—(Watson.)

938. *Polygonum minus*, Huds. Native. English type. Range 1.

C. Shore of Wastwater, half-way down, north-west side.— (Rev. A. Ley, B.)

W. Pond near the Windermere rifle-butts, with *Peplis*.—(B.)

939. *Polygonum aviculare*, L. (Knot Grass). Native. British type. Range 1. Roadsides and cultivated fields. Common, ascending to 250 yards in Troutbeck Valley; 330 yards.—(Watson.)

Var. *littorale* on the Furness shore at Kents Bank.

940. *Polygonum Raii*, Bab. Native. Maritime. British type. Range 1.

C. On the shore between Whitehaven and Workington. From this, sent by Lawson to Ray, the specific name was taken. Coast sand-hills south of Maryport.—(Professor Oliver.) Gathered by Miss Wilson at St. Bees.—(C. Bailey.) Shore at Seascale.—(Rev. A. Ley.) Shore between Braystones and Coulderton.—(W. Hodgson.)

941. *Polygonum Convolvulus*, L. (Black Bindweed). Colonist. British type. Range 1. Cultivated ground. Frequent; ascending to 250 yards at Windermere.

P. Fagopyrum, L. (Buckwheat). Sown as food for game; sometimes strays from cultivation.

944. *Rumex crispus*, L. Native. British type. Range 1. Roadsides and waste ground. Frequent; ascending to 300 yards at Shap.

Var. *trigranulatus* is plentiful on the shore at Arnside, Kents Bank, Flookborough, and Cark.

944. *Rumex aquaticus*, L. Native. Scottish type. Range 1.

W. Checked in Mr. Watson's Shap Catalogue, altitude 300 yards. I tried to find it there in 1883 without success.

945. *Rumex pratensis*, M. and K. Native. English type. Range 1.

C. St. Bees.—(Whitehaven Cat.)

W. Ascending Brantfell from Bowness.—(B.)

Rumex alpinus, L. (Butter Dockin ; Monk's Rhubarb). Alien. Cultivated as a pot-herb, and for wrapping round butter to keep it cool. Occasionally half-wild.

C. With *Imperatoria* by the roadside a mile from Mungrisedale towards Greystoke Park.—(Borrer.) Over Ullswater, about homesteads.—(W. Hodgson.) Hayton Castle.—(Rev. R. Wood.)

L. Orchard at Swarthmoor Hall, Ulverstone.—(B.)

947. *Rumex obtusifolius*, L. (Common Dock). Native. British type. Range 1-2. Everywhere common in ditches and by roadsides ; ascending to 500 yards at Hayes Water.

948. *Rumex nemorosus*, Schrad. Native. British type. Range 1. Woods and hedge-banks. Frequent in the low country; ascending to 250 yards in Lowther Woods and Troutbeck Valley.

Var. *sanguineus* is found occasionally in the Lamplugh district.—(W. Hodgson.)

948*. *Rumex conglomeratus*, Murr. Native. British type. Range 1. Mostly associated with the last, but at the Lakes much less common. Plentiful in the west of Cumberland about Aspatria and elsewhere.—(W. Hodgson.)

Rumex pulcher, L. A plant gathered by Lawson 'between the inn and smithy at Sir John Lowther's Newtown' is referred to this species by Professor Babington, but I suspect something else was really intended.

951. *Rumex Acetosa*, L. (Sour Dockin ; Common Sorrel). Native. British type. Range 1-4. Common in grassy places at all levels; ascending to 1000 yards on Helvellyn, and very high also on Skiddaw, Scawfell Pike, and Grassmoor.

R. scutatus, L., is cultivated as a pot-herb, and sometimes strays from gardens, as at Keswick and Allithwaite in Furness

952. *Rumex Acetosella*, L. (Sheep's Sorrel). Native. British type. Range 1-2. Dry hillsides and poor meadows. Frequent; up to 510 yards.—(Watson.)

953. *Oxyria reniformis*, Hook. (Mountain Sorrel). Native. Highland type. Range 1-3. High wet slate crags; not uncommon.

C. Piers Ghyll, Mickledore, Black Rocks of Great End, Sprinkling Tarn, Styhead Tarn, Wastwater Screes, and other cliffs of the Scawfell group of hills; first recorded by Wood. Honister Crag.—(B.) Vale of Newlands, as low as 500-600 feet.—(Watson.) Ashness Ghyll, above Barrow Falls.—(Winch, Watson.) West face of Glaramara.—(C. Bailey.) Scarf Gap over Ennerdale.—(Whitehaven Cat.)

W. Mardale, in a deep ghyll on the south-east side of the Dun Bull Inn.—(Hindson.) Red Screes, and near the head of the Stock Ghyll stream.—(T. J. Foggitt.) Striding Edge, and on the west side of Helvellyn over Dunmail Raise.—(W. Foggitt, J. C. Melvill, B.) Easedale Tarn and Dungeon Ghyll.—(B.) Tongue Ghyll waterfall, between Seat Sandal and Fairfield.—(A. W. Bennett.) Ravine that descends from Kirkstone Pass to Brothers Water.—(W. Hodgson.) Near Buckbarrow Well in Long Sleddale and on Little Harter Fell Crag; first recorded by Lawson. Head of Kentmere.—(J. Wilson.) Cliffs of High Street over Bleawater.—(Rev. A. Ley.)

ORDER THYMELÆACEÆ.

955. *Daphne Laureola*, L. (Spurge Laurel). Native. English type. Range 1.

C. Drawing with date March 1862, 'Rottington, near St. Bees.'—(Rooke's Flora.)

W. Rayrigg and Graythwaite woods, Windermere.—(F. Clowes.)

956. *Daphne Mezereum*, L. (Mezereon). Denizen. English type. Range 1.

C. In a wood near Wigton.—(W. B. Waterfall.)

W. With *D. Laureola* at Windermere.—(F. Clowes.) Whitbarrow.—(F. Clowes.)

L. Woods at Staveley and Rowdsey.—(Aiton.) Said to have been found in Colton Woods. A single plant in a thicket near Mansriggs Hall, doubtfully wild.—(Miss Hodgson.)

ORDER ASARACEÆ.

Asarum europæum, L. (Asarabacca). Alien.

C. Naturalised at Ormathwaite near Keswick.—(Winch.) In small quantity at Troutbeck in Borrowdale.—(C. Wright.) Hutton Woods.—(Cooke.)

W. Wood near Dalton Hall, Burton in Lonsdale.—(Hindson.) Kirkby Lonsdale, where it is gathered out of the woods for medical use.—(Dr. Bath, in Smith's Eng. Flora, ii. 342.) At Dale Head, Martindale; reported by Rev. W. Richardson.

ORDER EMPETRACEÆ.

960. *Empetrum nigrum*, L. (Crowberry). Native. Scottish type. Range 1-4. On all the higher fells. Descends to Grange Fell over Lodore, Walla Crag, Keswick, and Penrith Beacon Woods. Ascends over 1000 yards on Helvellyn and Scawfell Pikes, and nearly to the top of Skiddaw, Saddleback, Grisedale Pike, and Great Gable. Local name 'Lingberry,' the idea being that it is a berry-bearing form of *Calluna vulgaris*.

ORDER EUPHORBIACEÆ.

962. *Euphorbia Helioscopia*, L. (Wart Grass ; Great Spurge). Colonist. British type. Range 1. Frequent in cultivated ground ; ascending to 300 yards.—(Watson.)

? *Euphorbia Cyparissias*, L. Alien. An occasional escape from gardens.

C. Wall-top at Seatollar, Borrowdale.—(Rev. A. Ley!) Ulpha.—(C. Wright.)

W. Whitbarrow Fells, in plenty.—(Rev. W. H. Hawker.)

L. Jackland Tarn, Ulverstone.—(Miss Hodgson.)

969. *Euphorbia Paralias*, L. Native. Maritime. Atlantic type. Range 1.

C. On the shore at Harrington, Haverigg, and Flimby.—(Mr. Wood, J. Otley.) Millom.—(W. Hodgson.)

L. On the shore between Bardsea and Sandside, and by Park Head, Holker.—(Aiton.)

970. *Euphorbia portlandica*, L. Native. Maritime. Atlantic type. Range 1.

C. On the shore at Drigg and Braystones.—(W. Dickinson, M. Chambers.) Shore between Braystones and Coulderton, sparingly.—(W. Hodgson.)

971. *Euphorbia exigua*, L. Colonist. English type. Range 1. Cultivated fields. Rare. Whitehaven, Kendal, Arnside, etc. On the railway between Bullgill station and the first bridge over the river Ellen, westward, plentiful.—(W. Hodgson.)

972. *Euphorbia Peplus*, L. (Petty Spurge). Colonist. British type. Range 1. Cultivated fields. Frequent, up to 250 yards at Windermere village.

Euphorbia Lathyris, L. (Caper Spurge). Alien. Given in the Whitehaven list as occurring at Keswick (Fleswick intended?).

974. *Euphorbia amygdaloides*, L. (Wood Spurge). Native. English type. Range 1.

W. Woods near Milnthorpe.—(Miss Barnard.)

976. *Mercurialis perennis*, L. (Mercury). Native. British type. Range 1-2. Woods and hedge-banks. Common; ascending from shore-level at Conishead Priory to the limestone pavement of Whitbarrow, Hutton-Roof, and Farleton Knot, and to 400 yards in Troutbeck Valley.

Mercurialis annua, L. Alien?

C. In Eskdale at Dalegarth.—(J. Robson.)

ORDER URTICACEÆ.

978. *Urtica urens*, L. (Lesser Nettle). Native. British type. Range 1. Roadsides. Frequent, up to 300 yards at Shap and Keld, and in Borrowdale to Grange.

979. *Urtica dioica*, L. (Common Nettle). Native. British type. Range 1-2. Roadsides and waste ground; ascending to 500 yards at Styhead Pass, and on Coniston Old Man; 580 yards.—(Watson.)

982. *Parietaria diffusa*, Koch. (Wall Pellitory). Native. British type. Range 1.

C. On Crookdale Hall and Torpenhow Church.—(W. Dickinson, W. Hodgson.) Drawn by Rooke from the former station. A few plants grow on the south wall of Abbey Church, part of the ancient Abbey of Holm Cultram. —(W. Hodgson.)

L. Old limestone walls at Ulverstone, and foot of walls on the shore west of Humphrey Head.—(Miss Hodgson, B.) Walls of Furness Abbey, Bardsea Hall, and Cartmel Church. —(Aiton, C. Bailey.) Wraysholme Tower.—(C. J. Ashfield.)

Humulus Lupulus, L. (Wild Hop). Alien. Seen only in hedges near villages and farm-houses. Egremont, Gosforth, Keswick, Kirkby Lonsdale, Ennerdale, Watermillock, Cark, Allithwaite, etc.

984. *Ulmus montana*, Smith (Wych Elm). Native. British type. Range 1-2. One of the most frequent trees of the Lake woods. Ascends to 320 yards on the hills between Keswick and Thirlmere, and 400 yards in Hag Ghyll, Troutbeck, where it is the first tree one comes upon descending from High Street by the Roman road.

Ulmus suberosa, Ehrh. (Common Elm). Alien. Frequent in hedgerows in the low country and planted woods, but none here seen looking really wild. There are fine trees of typical *campestris* in Dalemain Park and Bowness Woods, of *major* in Dalemain Park and going out of Ambleside by the Kirkstone road, and I have a note of *glabra* as seen at Levens Bridge near Milnthorpe, and of *stricta* a little out of Troutbeck village towards Bowness.

ORDER AMENTIFERÆ.

988. *Quercus Robur*, L. (Oak). Native. British type. Range 1-2. Everywhere common in woods, and amongst the crags of the lower zone, both *pedunculata* and *sessiliflora*. Ascends from the shore cliffs at Humphrey Head to 350 yards on Grisedale Pikes, and the hills between Rosthwaite and Watendlath. Fine specimens in the park at Lowther and at Rydal Hall.—(W. Hodgson.)

Fagus sylvatica, L. (Beech). Alien. Common in parks and planted woods, as at Dalemain, and on the shore of Windermere at Bowness.

Castanea vulgaris, Lam. (Sweet Chestnut; Spanish Chestnut). Alien. Less frequent than the Beech, and seen only in parks and plantations, as at Furness Abbey and in Dalemain Park, Yewdale near Coniston, and by the river Crake near the mill. Two grand examples occur in Dalemain Park, nearly opposite Pooley Mill.

Carpinus Betulus, L. (Hornbeam). Alien. Not uncommon in hedgerows and plantations, as in Bowness Woods, Lowther Woods, and the head of Esthwaite Water between Colthouse and Hawkshead; also in the woods near Stybarrow Crag, Ullswater.

991. *Corylus Avellana*, L. (Hazel). Native. British type. Range 1-2. One of the commonest truly wild trees of the Lake district. Ascends to the limestone pavement of Hutton-Roof, Farleton Knot, and Whitbarrow; to 350 yards on the hills between Rosthwaite and Watendlath, and to 400 yards in Troutbeck Valley.

992. *Alnus glutinosa*, L. (Alder). Native. British type. Range 1. Sides of the lakes and streams. Common in the lower zone; ascending to 300 yards over Coniston, and 350 yards on Grisedale Pike.

993. *Betula alba*, L. (Birch). Native. British type. Range 1-2. Everywhere common in woods, and amongst the cliffs; ascending to 500 yards in Great Langdale, and noted by Watson at 600 yards over Thirlmere. Both the sub-species, *verrucosa* and *glutinosa* (*pubescens*), are frequent.

995. *Populus alba*, L. (White Poplar; Abele). Denizen. English type. Range 1. In many places by stream-sides in the low country.

997. *Populus tremula*, L. (Aspen). Native. British type. Range 1. Woods and hedges. Frequent, up to 300 yards in Troutbeck Valley, and 250 yards in Borrowdale on Castle Crag.

Var. *villosa* in the lane between Clifton and Great Strickland.

998. *Populus nigra*, L. (Black Poplar). Denizen. English type. Range 1. Stream-sides. Frequent; ascending in Borrowdale to Seatollar, and to 300 yards in Troutbeck Valley. A very fine tree on the river-bank at Eamont Bridge.

999. *Salix pentandra*, L. (Bay Willow). Native. Scottish type. Range 1. Lake-sides and hedges. Frequent throughout the district; ascending from shore-level at Meathop Moss to 300 yards at Shap.

1001. *Salix fragilis*, L. (Crack Willow). Native. British type. Range 1. Stream-sides. Frequent through the lower zone; ascending to 300 yards at Shap. Often planted, but truly wild in such places as the moss at Newton Regny near Penrith. There are some huge trees between Dacre Castle and Dalemain.—(W. Hodgson.)

Var. *decipiens* has been found by Rev. A. Ley near Foxfield railway junction.

1002. *Salix alba*, L. (White Willow). Denizen. British type. Range 1. Less common at the Lakes than the last, and often, perhaps always, planted. Ascends to 300 yards at Baldhow over Ullswater. Fine examples of this species at

Springkell, Aspatria. It also grows by the river Ellen, on the banks of which the var. *vitellina* is found in great abundance —(W. Hodgson.)

1004. *Salix triandra*, L. Native. English type. Range 1. Stream-sides in the low country. Not infrequent.

C. Banks of the Eden at Carlisle.—(Borrer.) Rosley.— (Rev. R. Wood.) Banks of the Eamont near Pooley Bridge. —(W. Hodgson, W. Foggitt.) Banks of the Ellen near Aspatria, and of the Caldew near Rose Castle.—(W. Hodgson.)

W. Stream-side between Bowness and Troutbeck, above the little waterfall.—(B.) Roadside near Hackthorpe and Newby, and banks of the Leith between Common Holme Bridge and Clibburn.—(B.)

1005. *Salix purpurea*, L. Native. British type. Range 1. Stream-sides and hedges in the low country.

C. Walk Mill and Whitehaven meadows.—(Whitehaven Cat.) Banks of Goldrill Beck and Ullswater shore at Skelley Nab, etc.—(W. Hodgson.) Banks of the Ellen below Aspatria, and of the Lowther at Askham.—(W. Hodgson.)

W. Banks of the Leith above Clibburn, and in the lane between Common Holme Bridge and Newby Head.—(B.)

L. By the canal feeder, Ulverstone.—(Miss Hodgson.) By a stream a little out of Hawkshead on the road to Ambleside. —(B.)

1006. *Salix rubra*, Huds. Native. English type. Range 1. Stream-sides in the low country. Rare.

C. Banks of the Eden at Carlisle.—(Borrer !) Westward. —(Rev. R. Wood.) Banks of the Ellen at Maryport and Aspatria, and of the Ehen at Kinniside.—(W. Hodgson.) Banks of the Eamont between Penrith and Pooley Bridge.— (W. Foggitt !)

1007. *Salix viminalis*, L. (Common Osier). Native. British type. Range 1. Stream-sides in the lower zone. Frequent; ascending to 250 yards in Troutbeck Valley, and 300 yards near Shap.

1008. *Salix Smithiana*, Willd. Native. English type. Range 1. Stream-sides and damp woods. Frequent; ascending from shore-level at Ulverstone to 300 yards over Coniston.

Var. *rugosa* was found by Miss Hodgson at Ulverstone, and var. *ferruginea* by Winch by the Lune at Kirkby Lonsdale. Mr. Hodgson includes *acuminata*, but not *Smithiana*, in his Ullswater list, and the Rev. R. Wood reports it from Westward. Smith's *acuminata* is the present species in part, but *S. dasyclados* is distinct.

1010. *Salix cinerea*, L. (Grey Sallow). Native. British type. Range 1-2. Woods and hedges. Common; ascending to 300 yards in Borrowdale, to the limestone pavement of Hutton-Roof, to 350 yards.—(Watson.)

Vars. *aquatica* and *oleifolia* both occur frequently.

1011. *Salix aurita*, L. Native. British type. Range 1-2. Swamps. Frequent; ascending to 300 yards over Stock Ghyll; 370 yards.—(Watson.)

1012. *Salix caprea*, L. (Great Sallow). Native. British type. Range 1-2. Woods and hedges. Common; ascending to 400 yards on the Stake Pass, and to the limestone pavement of Hutton-Roof.

1013. *Salix nigricans*, Fries. Native. Scottish type. Range 1.

C. Stream-sides round Ullswater, frequent.—(W. Hodgson.) In the moss at Newton Regny.—(B.) About Aspatria and Oughterside in the Ellen valley, fairly abundant.—(W. Hodgson.)

W. Banks of the Leith between Common Holme Bridge and Clibburn.—(B.) In a swamp by the roadside south of Witherslack Hall.—(B.)

1014. *Salix laurina*, Smith.   Native.   Scottish type. Range 1.

W. Banks of the Lune above the bridge at Kirkby Lonsdale (*S. tenuifolia*, Engl. Bot. tab. 2186).—(Borrer.)

1015. *Salix phylicifolia*, L.   Native.   Scottish type. Range 1-3.

C. Westward.—(Rev. R. Wood.)

W. Mountain rocks of Dollywagon Pike, Helvellyn, 2500 feet.—(Rev. A. Ley!) Stream-side below the Red Tarn, 1600-1700 feet.—(W. Hodgson!) Crosby Ravensworth.— (Watson.) River-side near the bridge at Kirkby Lonsdale.— (Borrer!)

1016. *Salix ambigua*, Ehrh.   Native.   British type. Range 1.

C. Bennet Head, Ullswater.—(W. Hodgson.) Westward. —(Rev. R. Wood.)

W. In the lane between Common Holme Bridge and Newby Head, associated with *S. aurita* and *repens*, between which it is no doubt a hybrid.—(B.)

1017. *Salix repens*, L.   Native.   British type.   Range 1-2. Dry heaths and sandhills; ascending from shore-level on Walney Island to 550 yards.—(Watson.) Mr. W. Hodgson sends characteristic specimens of var. *argentea* from Workington.

1022. *Salix Lapponum*, L.   Native.   Highland type. Range 3.

W. On the cliffs of Catchedicam, Helvellyn.—(J. Backhouse, Bolton King.)

Salix reticulata, L. Recorded by Mr. Sidebotham, Phytol. ii. 316, from a hill over Brothers Water, but I believe the true plant is quite confined to the Scottish Highlands. Mentioned in old lists as occurring in Greystoke Park, but this is, to say the least of it, doubtful.—(W. Hodgson.)

1027. *Salix herbacea*, L. Native. Highland type. Range 4. On the peaks of all the highest slate hills. This and *Carex rigida* are the two most Arctic plants of the Lake flora. I have a note of its having been found on Helvellyn, Glaramara, Saddleback, Skiddaw, Grassmoor, Hobcarten Fell, Grisedale Pike, Red Pike, Pillar, Scawfell Pike, and Kidsty Pike. Drawn from The Pillar in Rooke's series. Its lowest limit seems to be 2500-2600 feet.

1028. *Myrica Gale*, L. (Sweet Gale). Native. British type. Range 1-2. Frequent in the hill swamps; ascending from sea-level at Ulverstone to 400 yards on the hills between Rosthwaite and Watendlath (Watson); and the Seathwaite hills (Miss Hodgson).

ORDER CONIFERÆ.

Pinus sylvestris, L. (Scotch Fir). Frequently planted, as are the Spruce and Larch, from sea-level up to 500-600 yards.

1030. *Juniperus communis*, L. (Juniper). Native. British type. Range 1-3. Frequent both upon the limestone and slate hills, from Humphrey Head to 550 yards on Great Gable, 600 yards on Haystacks, and 800-900 yards on Helvellyn, Grisedale Pike, and Scawfell Pike.

Var. *nana*, dwarf and appressed to the rocks, grows in plenty, with Yew, on the limestone pavement of Whitbarrow, Farleton Knot, and Hutton-Roof.

1031. *Taxus baccata*, L. (Yew). Native. English type. Range 1-2. Frequent both on the slate and limestone crags; ascending from shore-level at Humphrey Head to the limestone pavement of Hutton-Roof and Whitbarrow, and to 500 yards on Grange Fell, over Castle Crag in Borrowdale. The finest trees in the district were those commemorated by Wordsworth in Lorton Vale and Borrowdale, one near a farm-house in Yewdale, north of Coniston, and those in the churchyard at Patterdale; but last winter the finest tree in Patterdale was blown down, and the group in Borrowdale completely wrecked. Many hills and cliffs are named after it,—Yewdale Crag near Coniston, Yew Crag below Seatollar and at Ullswater, and Yewbarrow at the head of Wastdale, near Grange-over-Sands and above the Winster valley opposite Whitbarrow.

ORDER ORCHIDACEÆ.

1032. *Goodyera repens*, R., Br. Native. Scottish type. Range 1.

C. In a fir plantation near the Eden at Armathwaite, between Penrith and Carlisle.—(Dr. F. A. Lees.)

1033. *Spiranthes autumnalis*, Rich. (Lady's Tresses). Native. Xerophilous. English type. Range 1.

L. Pastures about Newton near Cartmel.—(Mr. Hall.) Field by the road out of Kents Bank going towards Grange. —(B.) On the limestone, common near Baycliff, south-west of Ulverstone.—(Rev. R. Rolleston). Near Grange-over-Sands. —(T. Gough.)

1036. *Neottia Nidus-avis*, Rich. (Bird's Nest Orchis). Native. British type. Range 1. Shaded woods. Rare.

C. Walla Crag Wood, Keswick.—(Watson, W. Matthews.) Woodhall, Westward, and Flimby Wood near Whitehaven.—

(W. Dickinson.) Drawing in Rooke's Flora, 'Woodhall, Cockermouth.'

W. Cunswick Wood and other places near Kendal; first recorded by Wilson. Rare at Windermere.—(F. Clowes.) Buckham Wood, Lowther.—(T. J. Woodward.)

L. Not uncommon about Newton near Cartmel.—(Mr. Hall.) Coniston.—(Linton.) Roudsea Wood near Haverthwaite.—(Aiton, Miss Hodgson.)

1037. *Listera cordata*, Br. (Mountain Twayblade). Native. Scottish type. Range 1-2. Moory places. Rare.

C. Castlerig Fell, Keswick.—(W. Dickinson.) On Mellbreak, over Crummock, 500 yards.—(Wilson Robinson.) Fisher Ghyll Place, above the ravine where *Pyrola secunda* grows.— (Jos. Woods.) Ravenglass.— (Rooke's Flora.) Caldbeck Common.—(Cooke.)

W. On Cockley Fell, in Long Sleddale, and on the moors between Kendal and Shap.—(J. Wilson.) Moist moors near Kendal, not common.—(T. Gough.)

L. One place on Coniston Fell.—(Linton.)

1038. *Listera ovata*, Br. (Great Twayblade). Native. British type. Range 1. Woods and meadows. Frequent, up to 300 yards.

1039. *Epipactis latifolia*, All. (Helleborine). Native. British type. Range 1. Not infrequent in woods, from the shores of Derwentwater, Bassenthwaite, and Coniston Water, up to the limestone pavement of Whitbarrow, Hutton-Roof, and Farleton. A plant reported by Lawson from Abbot's Wood Close, near Great Strickland, is referred by Professor Babington to *E. ovalis*.

1040. *Epipactis palustris*, Sw. Native. English type. Range 1. Swamps. Rare.

C. In the moss at Newton Regny near Penrith.—(W. Hodgson, J. G. Goodchild.) Isell Woods near Bassenthwaite Lake.—(W. Dickinson.)

W. Cunswick Tarn and other marshes near Kendal.—(J. Wilson, T. Gough.) On Whitbarrow.—(W. Foggitt.) At Broadfell, north of Orton, with *Bartsia alpina*.—(T. Gough.)

L. Once seen near the lake at Coniston.—(Miss Beever.) Grange-over-Sands.—(I. Hindson, W. Foggitt.)

Cephalanthera grandiflora, Bab.

C. By the Bleng river above Gosforth.—(J. Robson.) Drawn by Rooke from this station.

W. In the Lowther woods near Askham.—(T. Lawson.) I think these will prove to belong to *C. ensifolia*.

1042. *Cephalanthera ensifolia*, Rich. Native. Xerophilous. English type. Range 1. Limestone woods. Very rare.

C. 'Workington Hall Park.'—(Rooke's Flora.)

W. Lowther Woods.—(T. J. Woodward.) Barrowfield Wood near Kendal.—(T. Gough.) On Whitbarrow.—(F. Clowes.

L. Old Park Wood, near Copse Head Rocks, Cartmel.—(Aiton.) On Yewbarrow, over Grange-over-Sands.—(J. Sidebotham, T. Gough.)

1045. *Orchis Morio*, L. (Purple Meadow Orchis). Native. English type. Range 1. Meadows. Rare.

C. Gosforth Bottoms.—(J. Robson.) Drawn from this station in the Rooke collection.

L. Plumpton Meadows and Old Hall Meadow, Ulverstone.—(Miss Hodgson.)

1046. *Orchis mascula*, L. (Purple Wood Orchis). Native. British type. Range 1. Woods and thickets. Frequent in the lower zone.

1048. *Orchis ustulata*, L. Native. Germanic type. Range 1. Grassy places. Rare.

C. At Blindcrake, north of Cockermouth.—(Rev. J. Dodd.) Tallantire.—(Rev. R. Wood.) Woodhall, Keswick.—(Linton.) The 'Tarnities' pastures near Cockermouth, and at Raughton Head, Carlisle.—(W. Hodgson.) Stainton Banks.—(W. Duckworth.)

W. Abundant on heathy ground at Arnside Knot.—(J. C. Melvill, C. Bailey.) Drawn by Miss Wilson from Warcop.

1051. *Orchis pyramidalis*, L. Native. Germanic type. Range 1. Grassy places. Very rare.

C. Gosforth Bottoms.—(J. Robson.) Drawn from this station in Rooke's collection. Watendlath.—(Black's Guide.)

1052. *Orchis latifolia*, L. Native. British type. Range 1. Damp meadows. Rare.

C. Between Seascale and Gosforth.—(J. Robson.) Ullswater shore at Waterfoot, and Furness railway near Nethertown station.—(W. Hodgson.) Watendlath and Borrowdale.—(Black's Guide.) At Langwathby and Salkeld.—(J. Wilson.) Margin of Thirlmere near the bridge (*incarnata angustifolia*).—(W. Matthews.)

W. Whittington Moss near Kirkby Lonsdale.—(I. Hindson.) Marshy ground in Arnside Woods.—(J. C. Melvill.)

L. Meadows at Cark and Cartmel.—(C. Bailey.)

1053. *Orchis maculata*, L. (Spotted Orchis). Native. British type. Range 1-2. Common in swampy ground; from shore-level to 500 yards on Great End.—(Watson.)

1054. *Gymnadenia conopsea*, Br. (Fragrant Orchis). Native. British type. Range 1. Meadows and pastures. Frequent; from shore-level at Holker to 250 yards in Watendlath Valley.—(W. Foggitt.)

1055. *Habenaria bifolia*, Br. (Butterfly Orchis). Native. British type. Range 1-2. Woods and meadows. Frequent, up to 420 yards.—(Watson.) Of the sub-species, *chlorantha* is the most frequent at the Lakes; *bifolia* occurs on the shores of Windermere and Derwentwater, at Dunmail Raise, and in Patterdale and the Watendlath Valley, and on Arnside Knot.

1056. *Habenaria viridis*, Br. (Frog Orchis). Native. British type. Range 1-2. Meadows and pastures. Frequent, up to 420 yards.—(Watson.) On Skiddaw, Brantfell, Shap Fells, etc. Very abundant in cow-pastures at Acrewalls, Frizington, along with *Botrychium Lunaria*.—(W. Hodgson.)

1057. *Habenaria albida*, Br. Native. Scottish type. Range 1-2.

C. Hillside above Watendlath Tarn, 350 yards, and in Borrowdale about the Bowder Stone, Ashness, and Barrow; first recorded by Winch. Little Broughton.—(Wilson Robinson.)

W. Barrowfield Wood, Kendal.—(T. Gough.) Loughrigg Tarn; drawn by Miss Wilson.

L. Cockley Beck, and on the banks of the Duddon in many places about Seathwaite.—(Miss Hodgson.) High marshy ground at Coniston.—(Miss Beever.)

1059. *Herminium Monorchis*, Br.

C. Ehen Valley at Cleator.—(J. Robson.) No doubt a misnomer. A drawing in Rooke's Flora, 'Rock tops going to the Lighthouse, St. Bees, July 1862.'

1060. *Ophrys apifera*, Huds. (Bee Orchis). Native. Xerophilous. English type. Range 1.

C. Meadows round Caldbeck near Wigton.—(Linton.) I

once saw a solitary specimen of this plant in a meadow at Aspatria.—(W. Hodgson.) Near Skiddaw.—(Rev. R. Wood.)

1061. *Ophrys muscifera*, Huds. (Fly Orchis). Native. Xerophilous. English type. Range 1. Limestone woods. Rare.

W. In the lane between Holme Park and the Crag; also in the wood, pretty plentiful.—(T. Lawson.) Barrowfield Wood, and in the marl close near Brigstear Moss, Kendal.— (J. Wilson.) Middlebarrow Wood, Arnside.—(J. C. Melvill.) Arnside Knot.—(T. W. Gissing.)

L. Plumpton Woods near Ulverstone.—(T. J. Woodward.) Wood-sides and fields about Grange-over-Sands. Common. —(J. Sidebotham.) Hagg Hills near Dalton in Furness, and Roudsea Wood, Haverthwaite.—(Aiton, Miss Hodgson.)

1063. *Malaxis paludosa*, Sw. Native. British type. Range 1.

C. One plant at Wastwater, July 1868.—(Miss Edmonds.) Lodore.—(C. Lyell.) Spongy bog at the foot of Grassmoor. —(W. B. Waterfall.)

W. Swindale Moors, over Hawes Water.—(Watson.) Swamp between Sölva How and Easedale.—(F. C. Roper.)

L. Between Rusland Chapel and Thwaite Moss near Newby Bridge.—(Rev. Mr. Jackson.)

1065. *Cypripedium Calceolus*, L. (Lady's Slipper).

C. Formerly grew in Vale of Legberthwaite.—(C. Wright.) A single plant was gathered near Blennerhasset about twenty years ago, and shown to me by the finder.—(W. Hodgson.) Said to have been found also on Whitbarrow and in the north-west of High Furness.

ORDER IRIDACEÆ.

1067. *Iris Pseudacorus*, L. (Yellow Iris). Native. British type. Range 1. Swamps and stream-sides. Frequent from shore-level in Furness up to Buttermere village and Rosthwaite in Borrowdale.

Crocus vernus, All. (Purple Crocus). Alien. An occasional straggler from cultivation.

W. Banks of the Mint near Kendal.—(T. Gough.)

L. In a field at Dragley Beck near Ulverstone, and near Swarthmoor Hall.—(Aiton.)

ORDER AMARYLLIDACEÆ.

Narcissus biflorus, Curt. Alien.

C. In two or three fields above High Lodore.—(W. Foggitt.)

L. Field by Coniston Lake near Torver.—(Miss Beever.) In Furness, near old halls.—(Miss Hodgson.)

1072. *Narcissus Pseudo-narcissus*, L. (Daffodil). Native. English type. Range 1. Woods and meadows.

C. Abundant on the shore of Ullswater in Glencoin Park. —(W. Hodgson.) Plentiful in Ennerdale and the Duddon valley.—(Whitehaven Cat.) Banks of the Irt.—(J. Robson.)

W. Abundant at Miller's Bay, St. Catherine's Wood, and other places about Windermere.—(F. Clowes, Rev. A. Bloxam.) Near Skelwith Bridge and Loughrigg Tarn.—(W. H. Hills.) Banks of the Mint, Kendal.—(T. Gough.) Great Strickland.—(Lawson.) Hedge-bank near Arnside Tower.— (C. J. Ashfield.)

L. High Thwaite, Coniston, in a field called Smartfield.— (Miss Beever.) Abundant in many places in Furness.— (Aiton, Miss Hodgson.)

Galanthus nivalis, L. (Snowdrop). Alien. An occasional straggler from cultivation.

C. Shores of Ullswater.—(W. Hodgson.) Near Brougham Castle, and in a wood at Dalemain.—(Mrs. King.) At the junction of Cockshot brook with the river Ellen near Torpenhow.—(W. Hodgson.) Lorton.—(W. B. Waterfall.)

W. Field near Beck Mills, Kendal.—(T. Gough.)

L. Common in Furness near houses and old halls.—(Aiton, Miss Hodgson.)

Leucojum æstivum, L. (Snowflake). Alien.

W. In a small island in the river three miles south of Kendal in the dam of the gunpowder mill.—(T. Gough.)

ORDER LILIACEÆ.

Lilium Martagon, L. (Turk's-cap Lily). Alien.

C. A casual near the mill at Dalemain.—(W. Hodgson.) Roebeck and banks of the Caldew.—(W. Duckworth.)

Fritillaria Meleagris, L. (Fritillary). Alien.

C. A casual at Oldchurch, Ullswater.—(W. Hodgson.)

W. In a meadow on the right-hand side of the road beyond Troutbeck on the way to Ambleside.—(Rev. A. Bloxam.)

1081. *Allium Scorodoprasum*, L. Native. Intermediate type. Range 1. Meadows. Rare.

C. Derwent banks near Workington.—(Mr. Tweddle.) Householm Island, Ullswater, and by river Ellen from Aspatria to Bullgill; plentiful at the latter station.—(W. Hodgson.) In Borrowdale near Lodore.—(Sir T. Frankland.) Banks of the Ehen.—(Whitehaven Cat.) Banks of Ullswater.—(J. Woods.)

W. Troutbeck Holm near Great Strickland.—(T. Lawson.)

Lowther woods and pastures adjacent.—(T. J. Woodward.) Near Mint's Bridge, Kendal.—(T. Gough.) Drawn from Fox How by Miss Wilson.

L. Meadow bordering the estuary at Pool Bridge, Rusland, —(Rev. Mr. Jackson, Miss Hodgson.)

1082. *Allium oleraceum*, L. (Wild Garlic). Native. Germanic type. Range 1. Meadows. Rare.

C. Borders of Derwentwater.—(D. Turner.) Householm Island, head of Ullswater.—(Rev. J. E. Leefe.) Is not the former species intended ?

W. Rocks in Long Sleddale.—(Dr. R. Richardson.) Troutbeck Holm, Great Strickland.—(T. Lawson.) Seamew Crag, Windermere.—(D. Turner.) Drawn from Windermere by Miss Wilson.

1083. *Allium vineale*, L. Native. English type. Range 1.

C. Bearpot near Workington.—(W. Dickinson.) Plentiful by the river Ellen between Blennerhasset and Aspatria.—(W. Hodgson.) Stainton banks.—(W. Duckworth.)

1085. *Allium Schœnoprasum*, L. (Chives). Native. Local type. Range 1.

W. Wet limestone rocks above Ruswittle in Lyth, Milnthorpe.—(T. Gough.)

L. Near Dalton in Furness.—(C. Wright.) By a small rivulet called Chivey Syke, Cartmel Fell.—(Rev. Mr. Jackson.) Nearly extinct there now, but abundant higher up in a bog remote from the road.—(Miss Hodgson.)

1086. *Allium ursinum*, L. (Ramps ; Wood Garlic). Native. British type. Range 1. Woods and thickets. Common up to 300 yards.—(Watson.)

1087. *Gagea lutea*, Ker. Native. Intermediate type. Range 1.

C. Boggy slope behind Redhills limestone quarry near Penrith.—(Mrs. Frank King.)

W. Banks of the Rothay.—(Dr. Davy.) Near Kendal.—(W. Hudson.)

L. Wood at Greenside, Heversham.—(T. Gough.) Grange-over-Sands.—(Rev. H. Higgins.)

Ornithogalum umbellatum, L. (Star of Bethlehem). Alien.

C. Meadow below Aspatria. Introduced with rubbish from a garden in the village.—(W. Hodgson.)

W. Meadow between Ambleside and Waterhead.—(W. Foggitt.) In a field near Kirkby Lonsdale.—(I. Hindson.)

L. Orchards in Furness.—(Miss Hodgson.)

1093. *Hyacinthus nonscriptus*, L. (Wood Hyacinth). Native. British type. Range 1-2. Woods and hillsides. Common, up to 500 yards on Saddleback. Varieties with white and rose-coloured flowers occur near Ulverstone and elsewhere.

1095. *Narthecium ossifragum*, L. (Bog Asphodel). Native. British type. Range 1-3. Everywhere common in swamps; from shore-level at Meathop Moss in Furness to 600 yards on Great Gable, 650 yards at Sprinkling Tarn; 720 yards.—(Watson.)

Asparagus officinalis, L. Alien.

L. On the shore rocks at Grange-over-Sands.—(Miss Hodgson.) I saw one plant there on the railway embankment in 1882.

Ruscus aculeatus, L. (Butcher's Broom). Alien.

C. A casual near Watermillock House, Ullswater.—(W. Hodgson.)

1099. *Convallaria majalis*, L. (Lily of the Valley). Native.

Xerophilous. Germanic type. Range 1. Woods, mainly on the limestone. Locally abundant.

W. Cunswick Wood, Hellsfell Nab, and other woods near Kendal.—(J. Wilson, T. Gough.) Limestone pavement of the top of Whitbarrow and Farleton Knot.—(B.) Witherslack Park, and on the scar near Waterfall Bridge, Great Strickland.—(T. Lawson.) Stony woods at Levens.—(Martyn.) Frequent in the woods about Kirkby Lonsdale.—(Hindson.) Covers acres in Middlebarrow Wood, Arnside.—(J. C. Melvill, B.)

L. Pull-wyke, Windermere.—(H. E. Smith.) Drawn from Skelwith by Miss Wilson. Isles of Windermere, now almost extinct.—(F. Clowes.) Abundant in Roudsea Woods near Haverthwaite.—(F. Clowes.) Hagg Hills, Dalton, and in Waitham Woods, Cartmel.—(Aiton.) Near the park northwest of Dalton.—(Mrs. Hart Jackson.)

1101. *Polygonatum multiflorum*, All. (Solomon's Seal). Native. English type. Range 1.

C. Near Dalegarth in Eskdale, in a coppice near the bridge.—(J. Robson.) In Castlehead Wood, Keswick, and in Borrowdale near Grange.—(Watson.) St. John's great wood, Keswick.—(Whitehaven Cat.)

W. In Rydal Park, near the house. Kendal, near farmhouses, not common.—(T. Gough.) Lowther Park.—(Rev. A. Ley.)

L. Abundant in Graythwaite Woods near lake-side, Windermere.—(F. Clowes.) Wood about Bigland Hall and Holker in Cartmel.—(Rev. Mr. Jackson.)

1102. *Polygonatum officinale*, All. Native. Xerophilous. English type. Range 1.

W. Barrowfield Wood near Kendal.—(T. Gough.) In the crevices of the limestone pavement of Farleton Knot.—(B.)

1103. *Paris quadrifolia*, L. (Herb Paris). Native. British type. Range 1. Frequent in woods and thickets in the lower zone; up to 300 yards at Shap Abbey.—(Watson.) Seen with seven leaves near the Lowther at Askham.—(W. Hodgson.)

ORDER DIOSCOREACEÆ.

1104. *Tamus communis*, L. (Black Bryony). Native. English type. Range 1. Frequent in woods and hedges in Furness and round Windermere, and about Arnside, Kendal, and Kirkby Lonsdale, but not seen about Penrith, Shap, Ullswater, or Keswick.

ORDER MELANTHIACEÆ.

1105. *Colchicum autumnale*, L. (Autumnal Crocus). Native. English type. Range 1. Damp meadows. Rare.

C. Swampy meadow below Blennerhasset (1861), since drained and brought under tillage; plant now probably extinct.—(W. Hodgson.)

W. At Greenside near Haversham, and at the junction of the Mint and Kent near Kendal.—(T. Gough.) Near Deepthwaite in Stainton and near the national school, Kirkby Lonsdale.—(I. Hindson.)

L. A little below Newby Bridge, on the left-hand side of the road to Ulverstone.—(J. Woods.)

ORDER HYDROCHARIDACEÆ.

Elodea canadensis, Rich. (American Waterweed). Alien. Now established both in Grasmere and Windermere.

Stratiotes aloides, L. (Water Soldier). Alien.

C. In Ennerdale Lake near Smithy Beck, 1852.—(J. Robson.) Sought there in vain by Mr. Hodgson.

ORDER ALISMACEÆ.

1109. *Alisma Plantago*, L. (Water Plantain). Native. British type. Range 1. Ponds and ditches. Frequent in the low country; ascending in Patterdale to Brothers Water, 250 yards.—(W. Hodgson.)

1110. *Alisma ranunculoides*, L. Native. British type. Range 1. Ponds and ditches. Not infrequent.

C. Eskmeals, Ravenglass.—(W. Dickinson.) Aspatria Moss.—(Rev. J. Dodd.) Braystones Tarn.—(J. Robson.) Edges of mossy pools at Dubmill near Allonby.—(W. Hodgson.)

W. Near the Cloven Stone on Great Strickland Moor.— (T. Lawson.) Plentiful in Windermere. — (F. Clowes.) Ponds by the railway side below Middlebarrow Wood, Arnside.—(I. Hindson, B.) Pond between Witherslack and Townend.—(B.)

L. Ditches at Goldmire near Dalton, and bogs near Roudsea Wood.—(Atkinson, Aiton.) Peat-ditches at Plumpton.— (Miss Hodgson.) Drawn from Wray by Miss Wilson.

1111. *Alisma natans*, L. Native. Local type. Range 1.

C. Derwentwater.—(Right Hon. C. Greville.) Braystones Tarn.—(J. Robson.)

L. Coniston.—(Miss Beever.) Confirmation wanted.

1113. *Sagittaria sagittifolia*, L. (Arrow Head). Native. English type. Range 1.

C. Ditches at Braystones near Beckermet.—(J. Robson.) Drawn from this station by Mr. Rooke.

L. Furness, frequent in bogs, ditches, and pools.—(Aiton.) I have not met with it, and it is not in Miss Hodgson's list.

1114. Butomus umbellatus, L. (Flowering Rush). Native. English type. Range 1.

C. In a pond at Irton.—(J. Robson.) River Wampool near Kirkbride.—(Rev. R. Wood.) In a pool near Lazonby Bridge, with *Elodea canadensis* on the north side of the river. —(W. Hodgson.)

1115. Triglochin maritimum, L. Native. Maritime. British type. Range 1. Frequent in salt-marshes; Allonby, Workington, Ulverstone, Cark, Flookborough, Milnthorpe, etc.

1116. Triglochin palustre, L. (Marsh Arrow Grass). Native. British type. Range 1-2. Frequent in swamps; ascending from the shore marshes at Flookborough to 400 yards on Coniston Old Man; 450 yards.—(Watson.)

ORDER POTAMACEÆ.

1118. Potamogeton densus, L. Native. English type. Range 1.

C. Allerby mill-race, near Maryport, where it is plentiful enough to be a nuisance to the miller in the autumn season. —(W. Hodgson.) In the Lowther at Askham.—(W. Hodgson.) Near Aspatria.—(Rev. R. Wood.)

W. Ditches by the stream at the north end of Shap village, with *Chara hispida*, 340 yards.—(Watson, B.) Plentiful in Leith about the bridge at Clibburn.—(B.)

1119. Potamogeton pectinatus, L. Native. British type. Range 1.

C. Bassenthwaite Lake.—(W. Dickinson.)

1121. Potamogeton pusillus, L. Native. British type. Range 1.

C. In Calder Ghylls near Ponsonby. — (J. Robson.) Plentiful in the old reservoir, Maryport, and in the mossy brook near Dubmill. —(W. Hodgson.) Westward.—(Rev. R. Wood.)

W. Ditches below Middlebarrow Wood, Arnside.—(C. Bailey, B.) In a pond in a hollow of the limestone on the top of Whitbarrow, 250 yards.—(B.)

1123. *Potamogeton obtusifolius*, M. and K. Native. English type. Range 1.

C. Harras Moor, Whitehaven.—(Linton.)

W. At the north end of Grasmere.—(F. C. Roper!)

1124. *Potamogeton crispus*, L. Native. British type. Range 1.

C. In the Derwent at Keswick.—(Linton.) Mockerkin Tarn.—(W. B. Waterfall.) In the mill-race at Egremont.— (Whitehaven Cat.) In the Eamont at Dalemain, the Ellen at Aspatria, in the fish-pond at Brayton Hall, in the Eden at Caldew, and in the Lowther at Askham.—(W. Hodgson.)

W. Abundant in Windermere.—(F. Clowes.)

L. Included in Miss Hodgson's Furness list.

1125. *Potamogeton perfoliatus*, L. Native. British type. Range 1. Abundant in the large lakes; Bassenthwaite Lake, Derwentwater, Ullswater, Windermere, Coniston Water, etc.

1126. *Potamogeton lucens*, L. Native. English type. Range 1.

C. Calder Ghylls, Ponsonby.—(J. Robson.) North-west end of Derwentwater.—(C. Bailey.)

L. Coniston Lake.—(Miss Beever.)

1127. *Potamogeton prælongus*, Wulf. Native. Scottish type. Range 1-2.

W. In Angle Tarn, Place Fell, 1550 feet. — (Borrer.) Windermere, abundant about the Ferry and elsewhere; first recorded by Borrer. Very abundant in Rydal Water.—(J. C. Melvill.) At the north end of Grasmere.—(F. C. Roper.)

L. At the north end of Coniston Lake near Waterhead. —(B.)

1129. *Potamogeton heterophyllus*, Schreb. Native. British type. Range 1. Plentiful in Ullswater, Rydal Water, and Windermere, and in the reservoir of Workington water-works.

Var. *Zizii* has been gathered by C. Bailey at the north end of Derwentwater and of Coniston Water. *P. lanceolatus* is reported by Mr. Cooke from Bassenthwaite, but the naming will need verifying.

1131. *Potamogeton rufescens*, Schrad. Native. British type. Range 1.

C. Calder Ghylls, Ponsonby.—(J. Robson.) Derwentwater.—(C. Bailey.) Dubstangs, and in the dam at Dubmill, Allonby.—(W. Hodgson.)

1132. *Potamogeton natans*, L. Native. British type. Range 1-2. Ponds in the low country, up to the lower tarn at Watendlath, 250 yards. At double this altitude, towards the source of Airey Beck, over Dowthwaite Head. — (W. Hodgson.)

1132*. *Potamogeton polygonifolius*, Pourr. Native. British type. Range 1-2. Common in tarns and swamps; ascending to 600 yards on Haystacks, and 500 yards at Hayes Tarn and Angle Tarn.

Var. *pseudo-fluitans* (Syme) has been gathered by the Rev. F. J. A. Hort and Professor Oliver at Buttermere, and in Stony Tarn near Wastwater by the Rev. A. Ley.

1132*. *Potamogeton plantagineus*, Ducr. Native. British type. Range 1.

C. Swamps in the moss at Newton Regny, gathered by Mrs. King and myself in 1883. Mossy ditch near Hurrock Wood, on Ullswater.—(W. Hodgson.)

1135. *Ruppia maritima*, L. Native. Maritime. British type. Range 1.

C. At Cloffocks, near Workington.—(W. Dickinson.)
W. Shore ditches just out of Arnside towards Milnthorpe (var. *rostellata*).—(B.)

1136. *Zannichellia palustris*, L. Native. British type. Range 1.

C. In the bed of Goldrill Beck close to Ullswater.—(W. Hodgson.) Mill-race, Aspatria East Mill; also in the race at Allerby Mill with *Potamogeton densus*, at both stations plentiful.—(W. Hodgson.) In the river Ellen.—(Rev. R. Wood.)

1137. *Zostera marina*, L. Native. Maritime. British type. Range 1.

C. On the shore at Bootle, cast up with the tide.—(W. Dickinson.)

ORDER LEMNACEÆ.

1138. *Lemna minor*, L. (Duck Weed). Native. British type. Range 1. Frequent in ponds in the low country, up to 300 yards at Shap.—(Watson.)

1141. *Lemna trisulca*, L. Native. English type. Range 1.
C. Ditches at Gosforth.—(J. Robson.)

ORDER ARACEÆ.

1142. *Arum maculatum*, L. (Wake Robin). Native. English type. Range 1. Woods and hedge-banks. Frequent in the lower zone; up to 250 yards in Troutbeck Valley, and on Yew Crag over Ullswater.

ORDER TYPHACEÆ.

1144. *Sparganium minimum*, Fries. Native. British type. Range 1.

C. Pond near Hallbank, Aspatria. — (W. Hodgson.) Ditches at head of Derwentwater.—(J. Otley, B.) In Dubbeck, Cleator.—(J. Robson.) Shoulthwaite Moss, Thirlmere. —(W. Dickinson.) Mockerkin Tarn.—(Whitehaven Cat.) Eel Tarn, Wastwater (var. *affine*).—(Rev. A. Ley.)

W. Abundant in the small tarns round Windermere.— (F. Clowes.) Codale Tarn, Easdale.—(F. C. Roper.) Pond between Witherslack and Townend.—(B.) Ditch below Middlebarrow Wood, Arnside.—(C. Bailey.)

L. Coniston Lake.—(Miss Beever.) Tarn by the side of the road between Hawkshead and Coniston.—(W. Southall.)

1145. *Sparganium simplex*, Huds. Native. British type. Range 1. Ponds and ditches. Frequent in the lower zone; from shore-level at Humphrey Head to 250 yards on Whitbarrow.

1146. *Sparganium ramosum*, Huds. (Great Bur Reed). Native. British type. Range 1. Frequent in ponds and ditches in the low country, from Walney Island and Urswick Tarn up to Stock Beck, Kendal.

1147. *Typha latifolia*, L. (Reed-mace). Native. British type. Range 1. Lakes and ponds. Frequent at a low level. The natives erroneously call it 'bulrushes,' and sometimes 'candle-wicks.'—(W. Hodgson.)

1148. *Typha angustifolia*, L. Native. English type. Range 1.

W. Rydal Water.—(J. C. Melvill.)

L. Urswick Tarn, Ulverstone.—(Miss Hodgson.) Blelham Tarn.—(W. H. Hills.)

ORDER JUNCACEÆ.

1150. *Juncus filiformis*, L. Native. Intermediate type. Range 1. Shores of the large lakes; locally plentiful. Bassenthwaite Lake, Lowes Water, Derwentwater, Thirlmere, Crummock Lake, head of Coniston Water, by Windermere at Bowness.

1151. *Juncus conglomeratus*, L. Native. British type. Range 1. Frequent in swamps, up to 560 yards.—(Watson.)

1151*. *Juncus effusus*, L (Common Rush; 'Sieves'). Native. British type. Range 1-3. Everywhere common in swamps, up to 500 yards on Styhead Pass, 600 yards on Coniston Old Man, 720 yards on Helvellyn.—(Watson.)

1151*. *Juncus diffusus*, Hoppe. Native. British type. Range 1.

L. Humphrey Head, at the bottom of the wood on the east side of the ridge.—(B.)

1152. *Juncus glaucus*, Sibth. Native. English type. Range 1. Roadsides and swamps. Frequent; up to 300 yards at Shap; 400 yards.—(Watson.)

1154. *Juncus maritimus*, Sm. Native. Maritime. British type. Range 1.

W. On the shore below Arnside Knot, and up the estuary towards Milnthorpe.—(Rev. J. H. Thompson, B.)

L. Furness shore, west of Humphrey Head, and plentiful at Cark and Flookborough.—(Miss Hodgson, B.) Salt-marsh at Ulverstone.—(Rev. A. Ley.)

1156. *Juncus acutiflorus*, Ehrh. (Closs). Native. British type. Range 1-2. Very common in swamps ; up to 500 yards on the Stake Pass and Styhead Pass ; 600 yards.—(Watson.)

1157. *Juncus lamprocarpus*, Ehrh. Native. British type. Range 1-2. Frequent about the tarns and hill-streamlets, up to 500 yards on Styhead Pass; 620 yards.—(Watson.) A variety approaching *nigritellus* was found by Mr. A. G. More on the shore of Coniston Lake.

1158. *Juncus obtusiflorus*, Ehrh. Native. English type. Range 1.

C. Hallbank Pond, Aspatria.—(W. Hodgson.)

L. Morecambe shore at Greenodd.—(Miss Hodgson!) The plant so called in Whitehaven Catalogue proved to be *lamprocarpus*.

1159. *Juncus supinus*, Moench. Native. British type. Range 1-2. Peaty swamps. Frequent; up to 500 yards on Styhead Pass, and by Hayes Water, and on Coniston Old Man.

Var. *subverticillatus* occurs in boggy ground near Braysteads over Ullswater.—(W. Hodgson.)

1160. *Juncus compressus*, Jacq. Native. British type. Range 1.

C. Kinniside Common, Ennerdale.—(J. Robson.) Arlecdon.—(W. Dickinson.)

1160*. *Juncus Gerardi*, Lois. Native. Maritime. British type. Range 1. Coast salt-marshes. Sellafield, Millom Marsh, Walney Island, Cark, Flookborough, Grange, Arnside, etc.

1162. *Juncus bufonius*, L. (Toad Rush). Native. British type. Range 1-3. Common in marshes up to Stake Pass; 650 yards.—(Watson.)

Var. *fasciculatus* in the salt-marshes at Flookborough.

1163. *Juncus squarrosus*, L. Native. British type. Range 1-4. Moors at all elevations. Frequent, up to 1000 yards on Helvellyn and Skiddaw.

1165. *Juncus trifidus*, L.

C. Hardknot, at the head of Eskdale.—(J. Robson.) No doubt a misnomer; *triglumis* probably intended.

1168. *Juncus triglumis*, L. Native. Highland type. Range 3-4. Hill swamps. Rare.

C. By Lowes Water.—(Rev. W. Wood.) On Hanging Knot at about 2000 feet.—(Professor Oliver.) On Saddleback by Scales Tarn.—(Bicheno.)

W. Striding-edge Cliffs, Helvellyn, 800-900 yards; first recorded by Bicheno. Fisherplace Ghyll, between Great Dodd and Helvellyn.—(J. Woods.) Mardale.—(Watson.) Fairfield.—(Winch.)

1169. *Luzula sylvatica*, Bich. (Great Wood-rush). Native. British type. Range 1-3. Woods and crags. Frequent; from the islands in Derwentwater and Windermere up to

800 yards on High Street, and over Sprinkling Tarn; 850 yards on Helvellyn.—(Watson.)

1170. *Luzula pilosa*, Willd. (Lesser Wood-rush). Native. British type. Range 1-2. Frequent in woods, up to 560 yards.—(Watson.) There are four distinct records for *L. Forsteri*, but I have no doubt the present species has been mistaken for it.

1172. *Luzula campestris*, Br. (Field Rush). Native. British type. Range 1-4. Grassy places; common at all elevations. Ascends to 950 yards.—(Watson.)

1173. *Luzula multiflora*, Lej. Native. British type. Range 1-4. Everywhere common in mossy places; from Newton Regny Moss up to 900-1000 yards on Skiddaw.

1175. *Luzula spicata*, DC. Native. Highland type. Range 3-4. Gathered on Fairfield by Woods, near the summit of Helvellyn by Sidebotham, and on Blake Fell, between Ennerdale and Lowes Water, by W. Dickinson.

ORDER CYPERACEÆ.

1178. *Cladium Mariscus*, R., Br. (Great Sedge). Native. English type. Range 1. Deep swampy ponds. Very rare.

C. In the moss at Newton Regny near Penrith, plentiful. —(W. Hodgson, B.)

W. Cunswick Tarn near Kendal; first recorded by Curtis, also by Otley.

1179. *Schœnus nigricans*, L. (Bog Rush). Native. British type. Range 1.

C. In Newton Moss with the *Cladium*.—(W. Hodgson, B.)

W. Frequent in bogs near Kirkby Lonsdale.—(Hindson.) A Westmoreland plant, issued as *S. ferrugineus* in Dickson's *fasciculus*, belongs to this species.

1180. *Rhyncospora alba*, Vahl. Native. British type. Range 1. Hill swamps. Not infrequent. Gatescarth Pass, Coniston Old Man, Blea Tarn in Little Langdale, Brantfell over Bowness, Brigstear Moss near Kendal, etc. A Kendal plant, referred by Withering to *Schœnus fuscus*, is probably var. *sordida*, which has been found lately by the Rev. A. Ley in Foulshaw Moss.

1182. *Blysmus compressus*, Panz. Native. English type. Range 1.

C. Waverton Wood near Wigton.—(Cooke.)

W. Stated by Hudson to be common in the county. By a stream by roadside a little out of Shap village on the west, 340 yards.—(Watson, B.) Banks of the river Leith just below the bridge at Clibburn.—(B.)

1183. *Blysmus rufus*, Link. Native. Maritime. Scottish type. Range 1.

C. On Whitrigg Flow, a little to the north of our limits. —(W. Hodgson.)

L. Sparingly in the salt-marsh south of Flookborough.—(B.)

1184. *Scirpus lacustris*, L. (Bulrush). Native. British type. Range 1. Frequent in the larger lakes. Derwentwater, Windermere, Grasmere, Rydal Water, Lowes Water, Urswick Tarn, etc.

1184 *b*. *Scirpus glaucus*, Smith. Native. Maritime. British type. Range 1.

L. Abundant in the marsh below Middlebarrow Wood, Arnside.—(B.)

1186. *Scirpus setaceus*, L. Native. British type. Range 1. Frequent in swamps, up to the Kirkstone, Kirkstone Pass, 500 yards.—(C. Bailey!) I have no record for *S. Savii*, which is almost certain to be found.

1190. *Scirpus maritimus*, L. Native. Maritime. British type. Range 1.

C. Salt-marshes at Allonby, Workington, and Sellafield.—(W. Dickinson, Whitehaven Cat.)

W. By the river Gilpin, between Gilpin's Bridge and Raven's Lodge.—(C. Bailey.)

L. Pool Bridge.—(Miss Hodgson.) Salt-marsh at Holker. —(Rev. A. Ley.)

1191. *Scirpus sylvaticus*, L. Native. British type. Range 1.

C. Islands of Derwentwater.—(Winch.) Banks of the Marron.—(W. Dickinson.) Banks of the Ehen.—(Whitehaven Cat.) By Dacre Beck below Thackthwaite, and in meadows below Ellen Villa, Aspatria.—(W. Hodgson.) Ullock near Dean.—(W. B. Waterfall.)

1192. *Scirpus palustris*, L. Native. British type. Range 1-2. Lakes and ponds. Frequent, up to 350 yards in Watendlath Valley.—(Watson.)

1194. *Scirpus multicaulis*, L. Native. British type. Range 1.

C. Shores of Ennerdale Lake.—(J. Robson.) Westward near Wigton.—(Rev. R. Wood.)

L. Islands of Windermere.—(W. Foggitt.)

1195. *Scirpus pauciflorus*, Lightf. Native. British type. Range 1-3.

C. Hill-sides over Derwentwater.—(Watson.) Murton Moss.—(W. Dickinson.) By the upper tarn at Watendlath.—(Watson.) Swamps in Mosedale by the path up Blacksail Pass.—(B.) Side woods in Ennerdale.—(W. Hodgson.)

W. Cliffs over Red Tarn, Helvellyn.—(W. Foggitt.) Shap. —(Watson.) On Fairfield.—(F. C. Roper.)

L. Shore marsh at Flookborough.—(B.) Shore marsh at Plumpton.—(Miss Hodgson.)

1196. *Scirpus cæspitosus*, L. Native. British type. Range 1-3. Frequent on damp fells and wet crags, from shore-level at Meathop Moss up to 800 yards on Helvellyn, and 900 yards on Scawfell Pike.

1197. *Scirpus acicularis*, L. Native. English type. Range 1.

C. Egremont.—(W. Dickinson.) Ennerdale Lake.— (J. Robson.) Black Moss.—(Whitehaven Cat.)

1198. *Scirpus fluitans*, L. Native. British type. Range 1. Swamps and ponds.

C. Cogra Moss, Lamplugh. — (W. Dickinson.) Now probably extinct. Wastdale Head, by the bridge midway between Ritson's Inn and the Lake.—(B.) Ditches on Salta Moss, Allonby.—(W. Hodgson.) In the river Ellen.—(Rev. R. Wood.)

W. Pond near the Windermere rifle-butts.—(B.)

L. Rills about Newton in Cartmel.—(Mr. Hall.)

1200. *Eriophorum vaginatum*, L. (Hare's-tail Cotton Grass). Native. British type. Range 1-4. Frequent in moory swamps at all elevations from shore-level up to 950 yards on Saddleback. 'After it turns white sheep are greedy after it, so it is called Moss-crops about Clibburn, Water Sleddale, and all places here.'—(Lawson.)

1201. *Eriophorum polystachyon*, L. (Common Cotton Grass). Native. British type. Range 1-3. Frequent in swamps at all levels, from Plumpton salt-marsh and Newton Regny Moss up nearly to the top of High Street, 800 yards. I have no record for *E. latifolium*, which is pretty sure to be found.

Kobresia caricina, Willd. Reported by Mr. W. Dickinson from Whillimoor near Whitehaven, but this is not a likely station. It is abundant in Teesdale, especially on Widdy Bank.

Carex simpliuscula, Wahl., founded on Westmoreland specimens gathered by Dawson Turner, has never been identified clearly.

1204. *Carex dioica*, L. Native. Scottish type. Range 1-3. Frequent in swamps, up to 500 yards on Kirkstone Pass (C. Bailey); and 650 yards at Sprinkling Tarn (Watson).

1205. *Carex pulicaris*, L. (Flea Sedge). Native. British type. Range 1-3. Frequent in swamps, up to 500 yards on Honister Crag, 600 yards on High Street; 650 yards.—(Watson.)

1209. *Carex stellulata*, Good. Native. British type. Range 1-3. Common in swamps at all levels, up to 800 yards on Helvellyn.—(W. Foggitt.)

1211. *Carex ovalis*, Good. Native. British type. Range 1-2. Frequent in swamps, up to 500 yards on Kirkstone Pass; 560 yards.—(Watson.)

1212. *Carex curta*, Good. Native. British type. Range 1-3. Frequent in swamps, up to 500 yards on Styhead Pass. Var. *vitilis* at nearly 3000 feet on Scawfell Pike.—(Rev. A. Ley!)

1213. *Carex elongata*, L. Native. English type. Range 1.
C. Snellings Mire.—(W. Dickinson.)

1214. *Carex remota*, L. Native. British type. Range 1. Damp woods. Frequent, up to 250 yards over Hawes Water. —(Watson.)

1215. *Carex axillaris*, Good. Native. English type. Range 1.
C. At Eskatt near Whitehaven.—(W. Hodgson!) Has been referred also to *C. Boenninghauseniana*, Weihe.

1217. *Carex intermedia*, Good. Native. English type. Range 1.
W. By the mill-dam at Kirkby Lonsdale.—(I. Hindson.)
L. Grassy slopes in front of Grange-over-Sands. — (Miss Hodgson.)

1218. *Carex arenaria*, L. Native. Maritime. British type. Range 1. Sands of the seashore at Harrington, Seascale, Walney Island, and Flookborough.

1220. *Carex muricata*, L. Native. British type. Range 1.
C. Banks of Ullswater at Airey Force and Waterfoot.— (W. Hodgson.) Roadside at Redhills near Penrith.—(B.) Roadside near Clibburn.—(W. Hodgson.)

1222. *Carex vulpina*, L. Native. British type. Range 1. Frequent in swamps in the low country, especially near the sea. Yeorton, Murton Moss, Humphrey Head, Walney Island, Pool Bridge, Cark, Arnside, etc.

1223. *Carex teretiuscula*, Good. Native. British type. Range 1.

C. Near Bennet Head, Ullswater, and in Newton Regny Moss near Penrith.—(W. Hodgson.)

L. Urswick Tarn, Ulverstone.—(Miss Hodgson.)

1224. *Carex paniculata*, L. Native. British type. Range 1.

C. In the moss at Newton Regny near Penrith.—(B.) Over Ullswater by the Blackdyke, Baldhow.—(W. Hodgson.) Salta and Eskatt Woods near Whitehaven; also in a boggy meadow near Ullock, Cockermouth.—(W. Hodgson.)

W. Frequent in bogs about Kirkby Lonsdale.—(I. Hindson.) Blind Tarn Moss.—(W. Matthews.)

1227. *Carex atrata*, L. Native. Highland type. Range 3.

W. Cliffs of the High Street and Helvellyn ranges.— (J. Backhouse.)

1228. *Carex vulgaris*, Fries. Native. British type. Range 1-3. Frequent in swamps, ascending from shore-level at Flookborough to 800 yards.—(Watson.)

1229. *Carex rigida*, Good. Highland type. Range 1-4. On the summit of all the higher peaks. I have notes of its occurrence on Helvellyn, Scawfell Pike, Great Gable, Green Gable, Great End, Grassmoor, Grisedale Pike, Glaramara, and Saddleback. It marks with *Salix herbacea* the bounds of the Mid-arctic zone.

1231. *Carex stricta*, Good. Native. English type. Range 1.

C. Bullgill Bridge.—(W. Dickinson.) Believed to have been seen at Derwentwater.—(W. Matthews.) Biglands near Wigton.—(Cooke.)

W. Shores of Rydal Water.—(W. Matthews.)

L. Windermere, especially on the islands near the Ferry. First recorded by Borrer.

1232. *Carex acuta*, L.   Native.   British type.   Range 1.

C. Blackdyke, Baldhow, and in a field below Bennet Head over Ullswater.—(W. Hodgson.)   Between Derwentwater and Bassenthwaite.—(W. Matthews.)

W. Marsh below Middlebarrow Wood, Arnside.—(B.)

L. Urswick Tarn near Ulverstone.—(Miss Hodgson.) In the Whitehaven Catalogue, *C. aquatilis*, Wahl., is reported from St. Bees, but no doubt the name is incorrect, as it is a high-mountain species.

1234. *Carex flava*, L.   Native.   British type.   Range 1-3. Frequent in swamps, ascending from shore-level at Flookborough to 500 yards on the Stake and Styhead Passes, 600 yards on High Street and Coniston Old Man; 650 yards.— (Watson.)

1235. *Carex extensa*, Good.   Native.   Maritime.   British type.   Range 1.   Shore marshes.

C. Banks of the Marron.—(W. Dickinson.)

W. Banks of the Leven estuary beneath Arnside Knot. —(C. Bailey, B.)

L. Foxfield Marsh, Duddon estuary.—(Miss Hodgson.) About the mouth of Cark Beck.—(B.)   On the shore at Ulverstone.—(Rev. A. Ley.)

1236. *Carex pallescens*, L.   Native.   British type.   Range 1. Frequent in damp meadows, up to 400 yards in Great Langdale.

1237. *Carex fulva*, Good.   Native.   British type.   Range 1-2. Frequent in swamps amongst the hills, from shore-level in Furness and Newton Regny Moss up to 500 yards at Hayes Water.

1238. *Carex distans*, L.   Native.   Maritime.   British type.

Range 1. Shore marshes at Cark, Flookborough, Humphrey Head, and Arnside.

1239. *Carex binervis*, Smith. Native. British type. Range 1-3. Frequent on damp moors, up to 800 yards on Scawfell over Sprinkling Tarn.—(Watson.)

1240. *Carex lævigata*, Smith. Native. British type. Range 1. Damp woods. Rare.

C. Over Ullswater at High Lowthwaite.—(W. Hodgson.)

W. In some places about Windermere.—(F. Clowes.)

1241. *Carex panicea*, L. Native. British type. Range 1-2. Common in swamps, from Barrow Island and Newton Regny Moss up to 630 yards at Sprinkling Tarn.—(Watson.)

1244. *Carex limosa*, L. Mentioned by Hudson as common in Westmoreland. It is said to grow at Biglands near Wigton by Mr. Cooke, but I have not seen a Lakeland specimen.

1246. *Carex strigosa*, Huds. Native. English type. Range 1.

C. Shaw Wood.—(Rev. R. Wood.)

1247. *Carex sylvatica*, Huds. Native. British type. Range 1. Frequent in woods in the lower zone, as at Egremont, Ulverstone, Grange-over-Sands, and Gowbarrow.

Carex pendula, Huds. Mentioned by Winch as frequent in Cumberland, but I know of no Lakeland station. At Whitefield House, Overwater, planted there by the late Mr. Gillbanks.—(W. Hodgson.)

1249. *Carex Pseudo-cyperus*, L. Native. English type. Range 1.

C. Moorside Woods near Whitehaven.—(W. Dickinson.)
W. Terrabank Tarn, Kirkby Lonsdale.—(I. Hindson.)

1250. *Carex glauca*, Scop. ('Pry'). Native. British type. Range 1-2. The commonest sedge of damp grassy places at the Lakes, from shore-level in Furness up to 500 yards on Catbells and the Stake Pass, 600 yards on High Street.

1251. *Carex præcox*, Jacq. Native. British type. Range 1. Dry banks. Frequent in the lower zone.

1252. *Carex pilulifera*, L. Native. British type. Range 1-3. Frequent in damp grassy places, up to 650 yards on Scawfell Pike, 850 yards on Helvellyn.—(Watson.)

1255. *Carex digitata*, L. Native. Xerophilous. English type. Range 1.
W. On the limestone at Hutton-Roof.—(I. Hindson.) Abundant a little beyond our bounds in Silverdale, where it was found by Mr. Sidebotham. Should be looked for about Grange and Cartmel.

1256. *Carex filiformis*, L. Native. Scottish type. Range 1. Swamps. Rare.
C. Workington.—(Mr. Tweddle.) In Newton Regny Moss near Penrith.—(W. Hodgson.)
W. Terrabank Tarn, Kirkby Lonsdale.—(I. Hindson.) Clibburn Moss.—(W. Hodgson.)

1257. *Carex hirta*, L. Native. British type. Range 1. Damp meadows. Frequent in the lower zone.

1258. *Carex ampullacea*, Good. Native. British type. Range 1-2. The commonest large sedge of the hill-swamps; ascending from shore-level at Arnside to 500 yards on the Stake Pass.

1259. *Carex vesicaria*, L. Native. British type. Range 1. With the last, but much less plentiful. Newfield in Furness, Windermere, Portinscale, Keswick, Derwentwater, Kirkby Lonsdale, etc.

1260. *Carex paludosa*, Good. Native. British type. Range 1.

C. By the river Ellen, near Allerby Mill, Maryport.—(W. Hodgson.)

W. In Winster Valley, in a swamp south of Witherslack Hall.—(B.)

L. In Furness at Tridley salt-marsh, and Urswick Tarn near Ulverstone.—(Miss Hodgson.)

1261. *Carex riparia*, Curt. Native. British type. Range 1.

C. Bolton, at the foot of Brocklebank Fell.—(Rev. R. Wood.)

W. Common in swamps about Kirkby Lonsdale.—(I. Hindson.)

ORDER GRAMINEÆ.

1269. *Phalaris arundinacea*, L. (Reed Grass). Native. British type. Range 1. Lake-sides and damp woods. Common in the lower zone, up to 250 yards in Troutbeck Valley and Patterdale; 300 yards.—(Watson.)

Var. *variegata*, roadside near Hawkshead, probably a garden outcast.

Phalaris canariensis, L. (Canary Grass). Alien. Occasionally seen in waste ground. Whitehaven, Kirkby Lonsdale, Holme Mill, lake-head of Windermere, etc. Usually found where bird-cages have been cleaned.

1271. *Anthoxanthum odoratum*, L. (Vernal Grass). Native.

British type. Range 1-3. Everywhere common in meadows, up to 900 yards on Helvellyn, and 850 yards on High Street.

1273. *Phleum pratense*, L. (Timothy Grass). Native. British type. Range 1. Meadows in the low country. Common, up to 300 yards at Mardale, Shap, and over Coniston.

1274. *Phleum arenarium*, L. Native. Maritime. English type. Range 1.

C. On the shore at Seascale.—(W. Foggitt.) On the shore at intervals from Silloth to Whitehaven.—(W. Hodgson.)

1278. *Alopecurus pratensis*, L. (Fox-tail Grass). Native. British type. Range 1. Common in meadows, up to 300 yards.

1279. *Alopecurus geniculatus*, L. Native. British type. Range 1-2. Common in ponds and ditches, up to 520 yards.—(Watson.)

Polypogon monspeliensis, Desf. Alien.

C. Ullswater shore at Floshgate; introduced with foreign corn.—(W. Hodgson.)

1287. *Milium effusum*, L. Native. British type. Range 1-2.

C. Flimby Wood, south of Maryport.—(W. Hodgson.) Between Rosthwaite and Seatollar, 1100 feet.—(Watson.)

W. Stock Ghyll, Ambleside.—(B.) In the Lowther Woods near Askham Bridge.—(W. Hodgson.)

1290. *Agrostis canina*, L. Native. British type. Range 1-2. Frequent in bogs, from shore-level at Ulverstone up to 600 yards on Coniston Old Man.

1291. *Agrostis vulgaris*, With. (Common Bent Grass). Native. British type. Range 1-4. Everywhere common in grassy places, from shore-level up to the top of High Street, Skiddaw, and Scawfell Pike.

Var. *pumila* is frequent by roads on the hills.

1292. *Agrostis alba*, L. Native. British type. Range 1. Swamps and grassy places. Frequent, from the Furness salt-marshes up to 610 yards.—(Watson.)

1293. *Ammophila arundinacea*, Host. (Marram). Native. Maritime. British type. Range 1.

C. Frequent in sandy ground along the coast through Cumberland.

L. In Furness at Roosebeck and Walney Island. Not seen about Grange and Arnside.

1294. *Arundo Phragmites*, L. (Great Reed). Native. British type. Range 1. Common in ditches and by lake-sides in the lower zone from the shore at Humphrey Head to 250 yards at Brothers Water.

Arundo Calamagrostis, L.

C. By the river Derwent.—(W. Dickinson.) This proved to be *Phalaris*.

W. Common in marshes at Kirkby Lonsdale.—(I. Hindson.) Confirmation needed.

1296. *Arundo Epigejos*, L. Native. English type. Range 1.

W. Limestone cliffs on the shore below Arnside Knot.—(B.)

L. Damp western slope of the hill near Elliscales Hall, Dalton in Furness.—(Dr. F. A. Lees.)

1299. *Sesleria caerulea*, Scop. Native. Xerophilous. Highland type. Range 1.

C. Abundant about Fell-end and the Side woods in Ennerdale; also at Moorside Hall, Arlecdon.—(W. Hodgson.) Coulderton.—(Whitehaven Cat.)

W. The commonest grass all over the limestone pavement of Shap Common, Whitbarrow, Farleton Knot, Hutton-Roof, and Arnside Knot, down to shore-level by the Milnthorpe estuary.

L. Plentiful about Grange-over-Sands and Humphrey Head.—(B.)

1300. *Aira cæspitosa*, L. (Bull Toppins). Native. British type. Range 1-3. Ditches and damp meadows. Common, from shore-level at Ulverstone up to 500 yards at Hayes Water, on Great Gable, and 600 yards on Helvellyn. A viviparous mountain form high up on Scawfell Pike.—(Rev. A. Ley.)

1302. *Aira flexuosa*, L. Native. British type. Range 1-4. Common on heaths at all elevations, up to 900 yards on Helvellyn, 950 yards on Great Gable, 1000 yards on Scawfell Pike.

1303. *Aira caryophyllea*, L. Native. British type. Range 1. Frequent on dry banks in the lower zone.

1304. *Aira præcox*, L. Native. British type. Range 1-2. Frequent on dry banks; up to 400 yards in Great Langdale, 500 yards on Kirkstone Pass.

Stipa pennata, L. (Feather Grass). Reported to have been found by T. Lawson and Dr. R. Richardson on limestone rocks in Long Sleddale, but often sought for there in vain by Mr. Gough and others.

1307. *Avena fatua*, L. Colonist. British type. Range 1. W. Corn-field near Barrow, Kirkby Lonsdale.—(I. Hindson.)

1308. *Avena strigosa*, Schreb. (Wild Oat). Colonist. British type. Range 1.

C. Ghyll near St. Bees.—(Rev. R. Wood.)

W. Frequent in corn-fields at Kirkby Lonsdale.—(I. Hindson.)

1309. *Avena pratensis*, L. Native. Xerophilous. British type. Range 1. Frequent in the limestone tracts; ascending to 300 yards near Shap.

1310. *Avena pubescens*, L. Native. British type. Range 1 Dry grassy places. Frequent; ascending from the coast at Aspatria to the limestone cliffs of Shap Common, 300 yards.

1311. *Avena flavescens*, L. Native. British type. Range 1. Common in meadows and pastures; ascending with the last to the limestone of Shap Common, 300 yards.

1312. *Arrhenatherum avenaceum*, Beauv. (Button Twitch). Native. British type. Range 1. Pastures and hedge-banks. Common, up to 250 yards at Lodore and Thrimby near Lowther.

1313. *Holcus lanatus*, L. (Yorkshire Fog). Native. British type. Range 1-2. Everywhere common in grassy places, up to 570 yards.—(Watson.)

1314. *Holcus mollis*, L. Native. British type. Range 1-2. Woods and grassy places. Frequent, from the shore at Flookborough to 400 yards in Great Langdale; 500 yards at Hayes Tarn.

1315. *Triodia decumbens*, Beauv. Native. British type. Range 1-2. Common in dry pastures; up to 550 yards on Styhead Pass, 600 yards on High Street and the limestone pavement of Hutton-Roof.

ORDER GRAMINEÆ. 229

1316. *Koeleria cristata*, Pers. Native. British type. Range 1.

C. On the shore from Cardurnock to St. Bees and Coulderton, sparingly.—(W. Hodgson.)

1317. *Melica uniflora*, Retz. Native. British type. Range 1. Frequent in woods and thickets, up to 250 yards on Castle Crag in Borrowdale, and 300 yards in Mardale.—(Watson.)

1318. *Melica nutans*, L. Native. Scottish type. Range 1.

C. Barrow Woods, Keswick.—(W. Foggitt.) Gowbarrow Woods, Ullswater.—(W. Hodgson.)

W. Near the waterfall on Scale How.—(W. Hodgson.) In some of the woods round Windermere.—(F. Clowes.) Mardale, 300 yards.—(Watson.)

1319. *Molinia cærulea*, Moench. Native. British type. Range 1-3. Frequent in swamps, from shore-level at Flookborough and Meathop up to 500 yards on Stake Pass, and at Hayes Water; 670 yards.—(Watson.)

1320. *Catabrosa aquatica*, Presl. Native. British type. Range 1.

C. On the shore at Coulderton, and in a brook in Lamplugh Hall cow-pastures.—(W. Hodgson, W. Dickinson.) Spring near Blencowe station.—(W. Hodgson.)

W. Not uncommon near Kirkby Lonsdale.—(I. Hindson.) Hillside west of Witherslack Hall.—(B.)

1321. *Glyceria aquatica*, Sm. Native. English type. Range 1.

C. Near Low Mill.—(Whitehaven Cat.)

1322. *Glyceria fluitans*, Br. (Flote Grass). Native. British type. Range 1-2. Ditches and ponds. Common, from shore-level in Furness up to 500 yards in the peaty streams south of Hayes Water.

1322 *b. Glyceria plicata*, Fries.   Native.   British type. Range 1.   Ponds and ditches in the lower zone.   Not infrequent.   Shap, Thrimby, Greystoke, Stainton, Newton Regny, Skelwith, Witherslack, Dalton in Furness, etc.

Sclerochloa maritima, Lindl.   I have no record for either this or *S. distans* within our bounds, and have sought for them in vain in Furness.   The former occurs in several places on the Cumberland shore of the Solway estuary.

1325. *Sclerochloa procumbens*, Beauv.   Native.   Maritime. English type.   Range 1.

C. On the railway bank near Brayton station.—(W. Hodgson, Rev. R. Wood.)   Also at Silloth, north of our limits.

1326. *Sclerochloa rigida*, Link.   Native.   British type. Range 1.

C. Brookfield near Wigton.—(Cooke.)

W. Walls round Windermere in several places.—(W. Foggitt.)

L. Grange-over-Sands.—(T. J. Foggitt.)

1327. *Sclerochloa loliacea*, Woods.   Native.   Maritime. English type.   Range 1.

C. Common about Whitehaven, whence it was first recorded by Mr. Newton in Ray's Synopsis.

1328. *Poa annua*, L.   Native.   British type.   Range 1-3. Common in grassy places, up to 900 yards on Helvellyn.

1330. *Poa alpina*, L.   Native.   Highland type.   Range 3.

C. Cliffs of the east face of Helvellyn, 800-900 yards.—

(Balfour.) Dollywagon Pike, Helvellyn.—(Rev. A. Ley.) On Skiddaw, found by Mr. W. Duckworth of Carlisle.— (W. Hodgson.)

1331. *Poa pratensis*, L. (Common Meadow Grass). Native. British type. Range 1-2. Everywhere common in grassy places up to 550 yards on Great Gable, 600 yards on High Street.

Var. *subcærulea*, which is specially mentioned by Sir J. E. Smith as a plant of Cumberland and Westmoreland, I have seen on walls by the shore at Flookborough.

1332. *Poa trivialis*, L. Native. British type. Range 1-3. Common in grassy places, up to 500 yards in Wastdale and at Hayes Water; 710 yards.—(Watson.)

1333. *Poa compressa*, L. Native. British type. Range 1.

C. Walls of Penrith Castle, and near Clifton Castle.—(B.)
W. Rocks at Colwith Force (var. *polynoda*).—(Borrer.) Not infrequent at Kirkby Lonsdale.—(I. Hindson.)
L. Walls at Grange-over-Sands, and on the railway bank near Cark station.—(B.)

1334. *Poa nemoralis*, L. Native. British type. Range 1.

C. Rocks above Lodore.—(W. Foggitt.) Stybarrow Crag, Ullswater, and by the road at Dalemain Park.—(W. Hodgson, B.) Abundant on walls of the cutting north of Eamont Bridge.—(W. Hodgson, B.) Clints Woods and Hale Woods. —(Whitehaven Cat.)
W. Walls at Clifton Castle and Great Strickland, and banks of the Leith above Clibburn.—(B.) Scandale, Stock Ghyll, Dungeon Ghyll, and other places about Ambleside.—(Borrer, B.) Common about Kirkby Lonsdale.—(I. Hindson.)
L. Walls between Lakeside station and Newby Bridge.— (B.)

1334*. *Poa Balfourii*, Parn. Native. Highland type. Range 3.

W. Dollywagon Pike, and on the north face of Fairfield at the head of Deepdale.—(Rev. A. Ley!)

1335. *Briza media*, L. (Quaking Grass; Dodderin Grass). Native. British type. Range 1-2. Common in grassy places, up to 500 yards at Hayes Water; 620 yards.—(Watson.)

1337. *Cynosurus cristatus*, L. (Dog's Tail Grass). Native. British type. Range 1-2. Common in grassy places, up to 500 yards on Kirkstone Pass; 580 yards.—(Watson.)

Cynosurus echinatus, L. Alien.

C. In a field of flax near the Keswick railway a mile from Penrith, 1883.—(B.)

1339. *Dactylis glomerata*, L. (Cock's Foot Grass). Native. British type. Range 1-2. Fields and dry crags. Common, ascending to the limestone pavement of Hutton-Roof; 350 yards.—(Watson.)

1340. *Festuca sciuroides*, Roth. Native. British type. Range 1.

C. Railway bank near Brayton station.—(W. Hodgson.)

W. Wall near Troutbeck Bridge.—(F. Clowes.) Wall at the bottom of Windermere village by the road to Bowness. —(B.)

L. Walls near the Ferry Inn.—(W. Foggitt.) Wall-top at Ash landing.—(J. H. Lewis.)

1340*. *Festuca pseudo-myurus*, Soy.-Will. (Capon's Tail). Native. English type. Range 1.

C. Top of a wall near the Castle gardens at Workington.—

ORDER GRAMINEÆ.

(W. Hodgson.) Coulderton.—(Whitehaven Cat.) Conglomerate rocks at the foot of Ullswater.—(W. Hodgson.)

W. Frequent about Kirkby Lonsdale.—(I. Hindson.)

1342. *Festuca ovina*, L. (Fescue Grass). Native. British type. Range 1-4. Woods and moors. Common at all elevations, ascending to the top of Great Gable, Grisedale Pike, Fairfield, Helvellyn, Skiddaw, and Scawfell Pike. The viviparous form is frequent, as low down as Stock Ghyll.

1343. *Festuca duriuscula*, L. Native. British type. Range 1-2. Common in pastures, up to 500 yards on Coniston Old Man and 800-900 yards on Helvellyn.

1344. *Festuca rubra*, L. Native. British type. Range 1.

C. On the shore at Redness Point near Whitehaven.—(W. Hodgson.)

W. L. In the shore marshes about Flookborough, Grange, and Arnside.—(B.)

1345. *Festuca sylvatica*, Vill. Native. Scottish type. Range 1. Shaded woods. Rare.

C. By the waterfall in Walla Crag Wood, Keswick.—(B.) Woodend.—(Whitehaven Cat.)

W. Rocky ravine near the Greenside lead mines.—(W. Hodgson.) Mardale.—(Watson.) About Stock Ghyll Force, first recorded by Sir T. Gage, and the lower fall at Rydal.—(B.)

1346. *Festuca elatior*, Auct. Native. British type. Range 1. Stream-sides and grassy places. Frequent, from shore-level at Arnside up to 300 yards at Shap.

1347. *Festuca pratensis*, Huds. Native. British type.

Range 1. Common in meadows, ascending to 300 yards at Shap.

Var. *loliacea* has been noted at Shap and near Arnside Tower.

1348. *Bromus giganteus*, L. Native. British type. Range 1. Everywhere common in woods, up to 250 yards at Lowther.

1349. *Bromus asper*, L. Native. British type. Range 1. In similar places to the last, with which it is usually associated, up to 250 yards in Lowther Woods.

1350. *Bromus sterilis*, L. Native. British type. Range 1. Wall-tops and hedge-banks. Not infrequent.

1353. *Bromus erectus*, Huds. Native. English type. Range 1.

C. Egremont.—(Whitehaven Cat.)

L. Believed to have been gathered near Furness Abbey.—(C. Bailey.)

1354. *Bromus secalinus*, L. Colonist. English type. Range 1.

C. Tallantire near Workington.—(Cooke.)

1355. *Bromus commutatus*, Schrad. Colonist. English type. Range 1.

C. Floshgate, Ullswater; introduced.—(W. Hodgson.) Roadside near Greystoke.—(B.)

W. In a field of clover and Italian rye-grass just south of Clifton village.—(B.)

1356. *Bromus mollis*, L. (Goose Corn). Native. British

type. Range 1-2. Common in meadows, up to 300 yards at Shap and 400 yards in Troutbeck Valley.

1357. *Brachypodium sylvaticum*, Beauv. Native. British type. Range 1. Common in woods and thickets, up to the limestone pavement of Whitbarrow and Hutton-Roof, and 300 yards in Great Langdale.

1359. *Triticum caninum*, Huds. Native. British type. Range 1-2. Hedge-banks and dry woods, up to 300 yards at Shap; 350 yards.—(Watson.)

1360. *Triticum repens*, L. (Twitch). Native. British type. Range 1. Hedge-banks and wall-tops; common, up to 300 yards at Shap.

Var. *littorale* on the shore at Walney Island (Rev. A. Ley!), and in the salt-marshes at Cark and below Humphrey Head (B.).

1361. *Triticum acutum*, DC. Native. Maritime. British type. Range 1.

C. On the beach at Flimby near Maryport.—(W. Hodgson!)
L. On the railway embankment west of Cark station; sparingly.—(B.)

1362. *Triticum junceum*, L. Native. Maritime. British type. Range 1.

C. On the coast at Allonby, Maryport, Lowea, Seascale, etc.—(W. Dickinson, W. Hodgson.)
L. On the Furness shore at Roosebeck.—(Miss Hodgson.)

1363. *Lolium perenne*, L. (Rye Grass). Native. British type. Range 1. Common in grassy places, up to 300 yards at Shap, 400 yards in Troutbeck Valley.

Lolium italicum, A. Br. (Italian Rye Grass). Alien. Now commonly planted in forage fields through the lower zone.

1364. *Lolium temulentum*, L. (Darnel). Colonist. British type. Range 1. Has been seen as a weed of cultivated ground at Whitehaven, Aspatria, Westward, Floshgate, Penrith, and Kirkby Lonsdale.

1366. *Hordeum sylvaticum*, Huds. Native. Xerophilous. English type. Range 1.

L. Limestone Wood behind Grayrigg villas, Grange-over-Sands.—(B.)

1368. *Hordeum murinum*, L. (Wild Barley). Native. English type. Range 1. Sandy ground near the coast. Very rare.

C. Flimby near Maryport.—(W. Dickinson.)

1369. *Hordeum maritimum*, With. Native. Maritime. English type. Range 1.

C. On the shore at Coulderton.—(W. Dickinson.) Occurs at Silloth, beyond our bounds.

1370. *Nardus stricta*, L. (Mat Grass). Native. British type. Range 1-4. Common on moors at all elevations, up to 900 yards on High Street, 1000 yards on Skiddaw.

1371. *Lepturus filiformis*, Trin. Native. Maritime. English type. Range 1.

C. Salt-marshes at Workington.—(Mr. Tweddle.)
L. Grange-over-Sands.—(W. Foggitt.)

ORDER FILICES.

1372. *Ceterach officinarum*, Willd. (Scale Fern; Rusty Back). Native. English type. Range 1. Walls and crevices of rocks, mainly in the limestone districts.

C. St. Bees, Sandwith, Mosser, and Gosforth.—(W. Dickinson, J. Robson.) Old wall at Cleator.—(Rev. F. Addison.) On Yew Crag and the banks of Aira Beck.—(Linton.)

W. Recorded by Lawson from the old bridge at Troutbeck, long since destroyed. In many places in the limestone country from Kendal by way of Whitbarrow to Arnside Knot.

L. Rocks of Humphrey Head and wall at Cark.—(Dr. Windsor, C. Bailey.) Old walls about Newlands.—(Miss Hodgson.) Brathay and Pull Bay, Windermere.—(J. Coward.) On a barn at Low Bank-ground, Coniston.—(Miss Beever.)

1373. *Woodsia ilvensis*, Br. Native. Highland type. Range 3. High slate crags. Very rare. Has been gathered on Scawfell and Helvellyn by Isaac Hudhart and F. Clowes; on Dove Crag, Fairfield, Hart Crag, Scandale Fell, the Red Screes, and at the head of Riggindale, Mardale, by J. Coward; and on the Hill Bell range by J. Backhouse. Recorded by T. Thompson in Phytologist, vol. i. p. 331, from the walls of the church at Crosby Ravensworth, which is a very unlikely station. A plant from Dove Crag, Fairfield, which has been referred to *W. alpina*, I have not seen.

1374. *Polypodium vulgare*, L. (Common Polypody). Native. British type. Range 1-3. Common on walls, rocks, and old trees. Ascends to the limestone pavement of Whitbarrow and Hutton-Roof, to 400 yards in Great Langdale, and noted by Watson up to 670 yards.

1375. *Polypodium Phegopteris*, L. (Beech Fern). Native. Scottish type. Range 1-3. Damp woods and hedge-banks.

Frequent throughout the district, from near shore-level in Roudsey Wood and the level of Windermere at Bowness up to 500 yards on Coniston Old Man, 550 yards on Haystacks, 800 yards on Scawfell Pike.

1376. *Polypodium Dryopteris*, L. (Oak Fern). Native. Scottish type. Range 1-3. In similar places to the last, ascending from shore-level at Ulverstone up to 800-900 yards on the precipices of the east face of Helvellyn.

1377. *Polypodium Robertianum*, Hoffm. Native. Xerophilous. English type. Range 1.

C. Reported from Scale Force, Buttermere.—(Rev. R. Wood.)

W. Crevices of the limestone rocks. Abundant on Whitbarrow, Hutton-Roof, Farleton Knot, Arnside Knot, and other hills of that neighbourhood.

L. In Furness on Hampsfield Fell.—(Miss Hodgson.)

1378. *Allosorus crispus*, Bernh. (Parsley Fern). Native. Highland type. Range 1-4. Everywhere on the slate hills amongst rocks and at the foot of walls, one of the most universal of the montane Lakeland plants. Ascends to 950 yards on Great Gable, over 1000 yards on Scawfell Pike, and to the summit of Helvellyn and Skiddaw. Descends nearly to the level of Ullswater on Hollin Fell and Birk Fell, and to the level of the stream in the Vale of St. John, at the foot of Great Dodd.

1379. *Cystopteris fragilis*, Bernh. (Bladder Fern). Native. British type. Range 1-3. Frequent on the drier rocks and scars, much less common amongst the slate than the limestone hills. Ascends to the limestone pavement of Whitbarrow and Hutton-Roof, and to 900 yards on Helvellyn. A form approaching *C. alpina* was found on Saddleback by Mr. S. F. Gray.

1381. *Cystopteris montana*, Link. Native. Highland type. Range 3. Found by Mr. Bolton King in 1880 on the rocks above the Red Tarn, Helvellyn. Reported also from Langdale and Brothers Water, but I have not seen specimens.

1382. *Polystichum Lonchitis*, Roth. (Holly Fern). Native. Highland type. Range 3. High slate crags. Rare. Has been found on the eastern face of Helvellyn and Fairfield by F. Clowes, I. Hudhart, and others; by the Rev. W. H. Hawker on Swarth Fell, by J. Coward over Bleawater, High Street, above Small Water, Nanbield Pass, and on Harter Fell above Mardale; by Mr. Cooke from Carrock Fell, and by J. Backhouse on the mountains over Hawes Water.

1383. *Polystichum aculeatum*, Roth. (Shield Fern). Native. British type. Range 1. Woods and hedge-banks, especially in the limestone tract. Frequent, up to 300 yards in Mardale.

1384. *Polystichum angulare*, Newm. Native. English type. Range 1. In similar places to the last. Not infrequent. Loughrigg Fell, Skelghyll Woods, Troutbeck Park, woods round Windermere, Nab Scar Woods, over Grasmere, Coniston, Whitbarrow, Grange-over-Sands, Arnside Knot, Ulverstone; abundant at Gleaston in Furness, etc.

1385. *Lastrea Thelypteris*, Presl. (Marsh Fern). Native. English type. Range 1. Wooded swamps. Very rare.

C. Irton woods near Gosforth.—(J. Robson.) Said to have been found about Ullswater at Glencoin and Bleawick, but not seen lately.

L. Roudsey Wood near Haverthwaite.—(Miss Hodgson.) Peat-mosses between Lakeside and Greenodd.—(J. Coward.)

1386. *Lastrea Oreopteris*, Presl. (Mountain Fern). Native. British type. Range 1-2. Woods in the hill-country, from the level of the large lakes up to 500 yards on Kirkstone Pass.

1387. *Lastrea Filix-mas*, Presl. (Male Fern). Native. British type. Range 1-3. Everywhere common in woods and amongst rocks, ascending to 650 yards on Scawfell Pike, 900 yards on Helvellyn.

Vars. *affinis, Borreri,* and *abbreviata* are all of frequent occurrence. A single plant of *L. remota*, which is just halfway between this and *spinulosa*, was found by I. Hudhart near Windermere in 1856, and Mr. Coward reports it from Brathay Woods.

1388. *Lastrea rigida*, Presl. Native. Xerophilous. Intermediate type. Range 1. Abundant, along with *Polypodium Robertianum*, in the crevices of the limestone pavement on Whitbarrow, Farleton Knot, Hutton-Roof, and Arnside Knot. It has been gathered by Mr. Mason in Lancashire near Grange.

1390. *Lastrea spinulosa*, Presl. Native. British type. Range 1. Shaded and damp woods in the lower zone. Not infrequent. Whitehaven, Ullswater, Derwentwater, Windermere, Coniston, Roudsey Wood, and in all the low mosses about Witherslack, Meathop, and Milnthorpe.

Var. *uliginosa* is recorded by Lowe as growing with *Osmunda*, in a wood on the west side of Derwentwater, and by W. Dickinson from Lowes Water.

1391. *Lastrea dilatata*, Presl. Native. British type. Range 1-3. Woods and thickets; everywhere common, from shore-level up to 600 yards on Coniston Old Man and 850 yards on Helvellyn.

Var. *dumetorum* has been found at Mardale, Tilberthwaite,

Elter Water, and on Loughrigg Fell, Bowfell, and Fairfield ; var. *collina* at Torver, Coniston, Elter Water, Loughrigg Fell over Grasmere, and at the foot of Gunner's How; and var. *alpina* by Mr. F. Clowes on the banks of Hawes Water.

1392. *Lastrea æmula*, Brack. Native. Atlantic type. Range 1. Shaded woods in the lower zone. Very rare.

C. St. Bees Head.—(Rev. G. Pinder.)
W. In a few places round Windermere.—(F. Clowes.)
L. Coniston ; rare, on very steep banks overhanging streams. —(Miss Beever.) Broughton in Furness.—(J. M. Barnes.) In some of the Kirkby Moor ghylls, on the side towards the Duddon.—(Miss Hodgson.) Dale Park, south of Esthwaite Water.—(J. Coward.)

1394. *Athyrium Filix-fæmina*, Roth. (Lady Fern). Native. British type. Range 1-2. Everywhere common in shaded woods and thickets, from shore-level up to 500 yards on Coniston Old Man ; 590 yards.—(Watson.)

Vars. *rhæticum* and *molle* are both frequent ; var. *latifolium*, Bab., was found by Miss Wright near Keswick.

1395. *Asplenium viride*, Huds. (Green Spleenwort). Native. Highland type. Range 1-3. Mountain rocks, especially on the limestone. Not infrequent. Carrock Fell, High Pike, Arnside Knot, Whitbarrow, Farleton Knot, Kendal Fell, Mardale, and also on Coniston Old Man, Wastwater Screes, Honister Crags, the Red Tarn and Deepdale precipices, rocks over Brothers Water, etc.

1396. *Asplenium Trichomanes*, L. (Maiden-Hair Spleenwort). Native. British type. Range 1-2. Walls and rocks, especially of limestone ; ascending to the top of Whitbarrow, Farleton Knot, and Hutton-Roof, and to 400 yards on the Troutbeck face of High Street.

Q

Var. *incisum* has been found in Borrowdale by Miss Wright, and at Lindale near Grange by Mr. A. Mason.

1397. *Asplenium marinum*, L. Native. Maritime. British type. Range 1.

C. Sea cliffs between Whitehaven and St. Bees.—(J. Robson, Rev. F. Addison.)

L. About the ruins of Peel Castle.—(Miss Hodgson.)

Asplenium fontanum, L. Recorded by Hudson from rocky places at Wybourn (meaning, doubtless, Wythburn), but this is probably a mistake.

1399. *Asplenium Adiantum-nigrum*, L. (Black Spleenwort). Native. British type. Range 1-2. Frequent on walls and the drier rocks, up to 400 yards.—(Watson.)

1400. *Asplenium Ruta-muraria*, L. (Wall Rue). Native. British type. Range 1-2. Frequent on walls and the drier rocks, from the shore at Humphrey Head up to 450 yards.—(Watson.) Limestone pavement of Whitbarrow, Farleton Knot, Hutton-Roof, and Shap Common.

1400*. *Asplenium germanicum*, Weiss. Native. Scottish type. Range 1-2. Slate rocks. Very rare. Has been found in Borrowdale by J. Flintoft and Miss Wright, and near Scawfell by Rev. W. H. Hawker and I. Hudhart; also on Skiddaw, Mr. Clowes tells me, and by J. Coward in 1853 in Little Langdale, and by Rev. R. Wood on Barfe Fell.

1401. *Asplenium septentrionale*, Hull. (Forked Spleenwort). Native. Scottish type. Range 1-2. Slate rocks. Rare. Has been found in several places in Borrowdale, and also in the Vale of Newlands, the Wastwater Screes, Honister Crag, Patterdale, the Red Screes, and over Crummock Water,

ORDER FILICES.

Grasmere, and Red Tarn, and in the Ambleside district near Dungeon Ghyll, and at the head of Little Langdale. First recorded by Hudson.

1402. *Scolopendrium vulgare*, Sym. (Hart's Tongue). Native. British type. Range 1. Rocky places. Frequent principally on the limestone, ascending to the summit of Farleton Knot and Hutton-Roof.

1403. *Blechnum boreale*, Sw. (Hard Fern). Native. British type. Range 1-3. Common everywhere in moory places from shore-level up to 550 yards on Styhead Pass, and 700 yards on Scawfell Pike.

1404. *Pteris aquilina*, L. (Common Brake). Native. British type. Range 1-2. Hillsides. Everywhere abundant, marking by its cessation the upper limit of Watson's Agrarian region. Noted at 550-600 yards on Glaramara, Great Robinson, Grisedale Pike, and over Kirkstone Pass and Buttermere Hawes.

1408. *Hymenophyllum Wilsoni*, Hook. (Filmy Fern). Native. Atlantic type. Range 1-2. On damp slate rocks. Frequent all through the central Lake district. The plant was first named by Ray from specimens gathered by Lawson and Newton on Wrynose. It grows at Dalegarth in Eskdale, in the Calder Ghylls at Ponsonby, in the Duddon Valley at Seathwaite, at High Stile, Scale Force, and other places round Buttermere and Crummock; in Ennerdale on Angler's Rock, on Wastdale Screes, Scawfell and Great End, and down Borrowdale to Lodore; on Skiddaw, Coniston Old Man; on Bowfell, Rossett Ghyll, Dungeon Ghyll, and other places in Langdale; at Wythburn, Rydal Head, Sölva How, Tongue Ghyll, and several places round Grasmere; over Ullswater on Stybarrow Crag, and at Glencoin about Hawes Water, and on Casterton Fell near Kirkby Lonsdale. Mr. Clowes and Miss Beever both think all the Lake filmy fern is this sub-species, none of it true *Tunbridgense.*

1409. *Osmunda regalis,* L. (Flowering Fern). Native. British type. Range 1. Once spread through the district in low damp woods, but now much dug up and destroyed. For instance at Colwith Force, where it once grew luxuriantly, I was grieved at my last visit in 1882 to find none left. On the Brathay, where it was once, Mr. Coward tells me, so plentiful as to interfere with the fishing, it is now scarce, and on Loughrigg, and on the shores of Grasmere and Rydal lakes, where it was once abundant, there is now none left. It was found not only round Derwentwater, Windermere, and Coniston Water, but also along the coast at Allonby, Seascale, Sellafield, Ulverstone, and Grange, and in many of the mosses about Kendal, Milnthorpe, and Kirkby Lonsdale.

1410. *Botrychium Lunaria,* Sw. (Moonwort). Native. British type. Range 1. Frequent on heathy pastures, from shore-level at St. Bees Head up to Gowbarrow Fells (W. Hodgson), and Watermillock, 300 yards. Two varieties, called by Ray 'Branched Moonwort,' and 'Cut-leaved Moonwort,' were gathered by Lawson at Great Strickland.

1411. *Ophioglossum vulgatum,* L. (Adder's Tongue). Native. British type. Range 1. Frequent in meadows in the lower zone, but easily overlooked.

ORDER LYCOPODIACEÆ.

1412. *Lycopodium clavatum,* L. (Stag's-horn Moss). Native. British type. Range 1-3. Frequent on the grassy hillsides, up to 600 yards on Grisedale Pike and Haystacks; 700 yards on High Street.

1413. *Lycopodium annotinum*, L. Native. Highland type. Range 2.

W. On the Langdale side of Bowfell in several places, about half a mile to the left of the Stake Pass, discovered by the Rev. R. Rolleston, and gathered also by Mr. Clowes on the opposite slope of Bowfell.

1414. *Lycopodium inundatum*, L. Native. British type. Range 1-2. Damp moors. Rare. Wastdale Head.—(W. Dickinson.)

C. Shores of Wastwater.—(S. P. Woodward.) In Borrowdale near the Bowder Stone.—(D. Turner, Winch.) Shoulthwaite Moss, Thirlmere.—(W. Dickinson.)

W. Blea Tarn in Little Langdale, and other places round Windermere.—(F. Clowes, W. Foggitt.) On Loughrigg, and near Fell Foot in Little Langdale.—(Miss Beever.) Foot of Red Screes.—(C. Bailey.) Foot of Long Sleddale.—(T. Lawson.)

1415. *Lycopodium alpinum*, L. (Alpine Club Moss). Native. Highland type. Range 1-4. Frequent on the grassy slopes of the slate hills. Descends to 150 yards at Lodore, and in the Cocker valley to the level of Buttermere. Ascends to 900-1000 yards on Skiddaw, and 1000 yards on Helvellyn.

1416. *Lycopodium Selago*, L. (Fir Club Moss). Native. British type. Range 1-4. Frequent in damp turfy places. Descends to the shore of Wastwater, and banks of the stream at Gatescarth. Ascends to 900 yards on High Street, 950 yards on Great Gable, 1000 yards on Helvellyn and Scawfell Pike.

1417. *Lycopodium selaginoides*, L. Native. Highland type. Range 1-3. Frequent in damp moory places, from the level of Wastwater, 200 feet, to 900 yards on the Striding-Edge cliffs.

1418. *Isoetes lacustris*, L. (Quillwort). Native. Highland type. Range 1-2. In all the large lakes, and in many of the mountain tarns. Wastwater, Ennerdale Lake, Crummock Lake, Derwentwater, Windermere, Ullswater, Hawes Water, Rydal Water, etc. Ascends to Stickle Tarn, 500 yards.— (J. C. Melvill.) Codale Tarn (W. Matthews), and Floutern Tarn (W. Wilson).

ORDER MARSILEACEÆ.

1419. *Pilularia globulifera*, L. (Pillwort). Native. British type. Range 1.

C. Pond at Nethertown.—(Whitehaven Cat.) Ennerdale Lake.—(J. Robson.)

ORDER EQUISETACEÆ.

1420. *Equisetum maximum*, Lam. (Great Mare's Tail). Native. English type. Range 1. Damp woods. Not infrequent. Flimby, Whitehaven, Salter Hall, Parton, banks of the Irt, Keswick, banks of the Eamont, etc.

1421. *Equisetum umbrosum*, Willd. Native. Scottish type. Range 1.

C. Eskatt Wood.—(Whitehaven Cat.) I fear a mistake for *E. maximum*, which Mr. Hodgson tells me grows there abundantly.

W. Swindale.—(J. Backhouse.)

1422. *Equisetum arvense*, L. (Common Mare's Tail; Paddock Pipes). Native. British type. Range 1. Common by roadsides, and in cultivated fields, up to 300 yards. —(Watson.)

1423. *Equisetum sylvaticum*, L. (Wood Mare's Tail). Native. British type. Range 1-2. Frequent in damp shaded woods. Ascends to Swindale and Watendlath, 350 yards (Watson), and to the copper mine on Coniston Old Man (Miss Beever).

1424. *Equisetum palustre*, L. (Bog Mare's Tail). Native. British type. Range 1-2. Lakes and stream-sides. Frequent, ascending to 450 yards.—(Watson.)

Var. *polystachyon* has been found by Mr. Clowes at Windermere, and by Mr. W. Hodgson near Winder railway station.

1425. *Equisetum limosum*, L. Native. British type. Range 1-2. Ponds, tarns, and lakes. Common. Ascends to Brothers Water, Watendlath Tarn, and a pond near the top of Whitbarrow, 250 yards; to Seathwaite Tarn, 400 yards. —(Mrs. Hodgson.)

1426. *Equisetum hyemale*, L. (Dutch Rush). Native. Scottish type. Range 1. Damp woods. Rare.

C. Near a colliery reservoir in Flimby Wood, and in boggy fields in Kinniside, Whitehaven.—(W. Hodgson.) Brayton Woods near Allonby.—(W. B. Waterfall.) Bolton, at the foot of Brocklebank Fell.—(Rev. R. Wood.)

W. By the stream between Shap and Anna Well, 300 yards. —(T. Lawson.) Swindale.—(J. Backhouse.) Kirkby Lonsdale.—(J. Hindson.)

L. Wooded ghyll near Penny Bridge, Ulverstone.—(Miss Hodgson.)

1428. *Equisetum variegatum*, W. and M. Native. Scottish type. Range 1.

C. Wastdale.—(N. J. Winch.)

W. Banks of a stream at Shap, 300 yards.—(Watson.)

L. On the Furness shore near the road to Holme Island, sparingly.—(Professor C. C. Babington.)

POSTSCRIPT.

Saussurea alpina has been discovered in Cumberland this summer (1884) by Mr. W. B. Waterfall near Floutern Tarn.

The plant recorded from Grange-over-Sands as *Diotis* proves to be *Filago germanica*.

Chrysocoma Linosyris. Through the kindness of Professor Babington I have seen a specimen gathered in Furness near Hampsfield by Mr. W. Nixon of Eccleriggs.

Agrostis nigra, With., has been found by the Rev. W. W. Newbould in Furness, near Grange-over-Sands.

Asplenium lanceolatum has been found on the Cumbrian side of the Duddon estuary near Millom, by the Rev. W. T. Baker.

INDEX OF SCIENTIFIC NAMES.

Acer, 58.
Aceraceæ, 58.
Achillea, 141.
Aconitum, 23.
Actæa, 23.
Adonis, 17.
Adoxa, 104.
Ægopodium, 107.
Æthusa, 109.
Agrimonia, 87.
Agrostis, 225.
Aira, 227.
Ajuga, 162.
Alchemilla, 88.
Alisma, 205.
Alismaceæ, 205.
Allium, 200.
Allosorus, 238.
Alnus, 187.
Alopecurus, 225.
Althæa, 55.
Amaryllidaceæ, 199.
Amentiferæ, 186.
Ammophila, 226.
Anagallis, 173.
Anchusa, 169.
Andromeda, 143.
Anemone, 17.
Angelica, 110.
Anthemis, 141.
Anthoxanthum, 224.
Anthriscus, 112.
Anthyllis, 67.
Antirrhinum, 156.
Apium, 106.
Apocynaceæ, 147.

Aquilegia, 22.
Arabis, 35.
Araceæ, 210.
Araliaceæ, 104.
Arctium, 129.
Arenaria, 50.
Armeria, 174.
Armoracia, 33.
Arrhenatherum, 128.
Artemisia, 134.
Arum, 210.
Arundo, 226.
Asaraceæ, 183.
Asarum, 183.
Asparagus, 202.
Asperula, 117.
Asplenium, 241.
Aster, 136.
Astragalus, 71.
Athyrium, 241.
Atriplex, 177.
Atropa, 151.
Avena, 227.

Ballota, 162.
Balsaminaceæ, 64.
Barbarea, 36.
Bartsia, 154.
Bellis, 139.
Berberaceæ, 23.
Berberis, 23.
Beta, 177.
Betula, 187.
Bidens, 133.
Blechnum, 243.
Blysmus, 215.

Boraginaceæ, 166.
Borago, 169.
Botrychium, 244.
Brachypodium, 235.
Brassica, 38.
Briza, 232.
Bromus, 234.
Bryonia, 96.
Bunium, 108.
Bupleurum, 108.
Butomus, 206.

CAKILE, 29.
Calamintha, 161.
Callitriche, 95.
Calluna, 143.
Caltha, 21.
Camelina, 34.
Campanula, 141.
Campanulaceæ, 141.
Caprifoliaceæ, 113.
Capsella, 30.
Cardamine, 34.
Carduus, 129.
Carex, 219.
Carlina, 131.
Carpinus, 187.
Carum, 107.
Caryophyllaceæ, 44.
Castanea, 187.
Catabrosa, 229.
Caucalis, 111.
Celastraceæ, 64.
Centaurea, 132.
Centranthus, 118.
Centunculus, 173.
Cephalanthera, 195.
Cerastium, 52.
Ceterach, 237.
Chærophyllum, 112.
Cheiranthus, 38.
Chelidonium, 26.
Chenopodiaceæ, 176.
Chenopodium, 176.
Chrysanthemum, 139.
Chrysocoma, 133, 248.
Chrysosplenium, 104.
Cichorium, 128.

Cicuta, 106.
Cineraria, 138.
Circæa, 93.
Cistaceæ, 40.
Cladium, 214.
Claytonia, 53.
Clematis, 15.
Cochlearia, 31.
Colchicum, 204.
Comarum, 78.
Compositæ, 120.
Coniferæ, 192.
Conium, 106.
Convallaria, 202.
Convolvulaceæ, 149.
Convolvulus, 149.
Cornaceæ, 105.
Cornus, 105.
Coronopus, 29.
Corydalis, 127.
Corylus, 187.
Cotyledon, 101.
Crambe, 29.
Crassulaceæ, 99.
Cratægus, 89.
Crepis, 123.
Crithmum, 110.
Crocus, 199.
Cruciferæ, 29.
Cucurbitaceæ, 96.
Cuscuta, 150.
Cynoglossum, 169.
Cynosurus, 232.
Cyperaceæ, 214.
Cypripedium, 198.
Cystopteris, 238.

DACTYLIS, 232.
Daphne, 182.
Daucus, 111.
Delphinium, 23.
Dianthus, 44.
Digitalis, 156.
Dioscoreaceæ, 204.
Diotis, 134, 248.
Dipsaceæ, 119.
Dipsacus, 119.
Doronicum, 138.

INDEX OF SCIENTIFIC NAMES.

Draba, 33.
Drosera, 43.
Droseraceæ, 43.
Dryas, 77.

ECHINOPHORA, 113.
Echium, 169.
Elodea, 204.
Empetraceæ, 183.
Empetrum, 183.
Epilobium, 91.
Epimedium, 24.
Epipactis, 194.
Equisetaceæ, 247.
Equisetum, 247.
Erica, 143.
Ericaceæ, 143.
Erigeron, 136.
Eriophorum, 217.
Erodium, 59.
Eryngium, 105.
Erysimum, 37.
Erythræa, 148.
Euonymus, 64.
Eupatorium, 133.
Euphorbia, 184.
Euphorbiaceæ, 184.
Euphrasia, 154.

FAGUS, 187.
Festuca, 232.
Filago, 135.
Filices, 237.
Fœniculum, 109.
Fragarea, 78.
Fraxinus, 147.
Fritillaria, 200.
Fumaria, 28.
Fumariaceæ, 27.

GAGEA, 201.
Galanthus, 200.
Galeopsis, 163.
Galium, 115.
Genista, 66.
Gentiana, 147.
Gentianaceæ, 147.
Geraniaceæ, 59.

Geranium, 60.
Geum, 76.
Glaucium, 26.
Glaux, 174.
Glyceria, 229.
Gnaphalium, 134.
Goodyera, 193.
Gramineæ, 224.
Grossulariaceæ, 97.
Gymnadenia, 196.

HABENARIA, 197.
Haloragiaceæ, 94.
Hedera, 105.
Helianthemum, 40.
Helleborus, 22.
Helminthia, 121.
Helosciadium, 106.
Heracleum, 111.
Herminium, 197.
Hesperis, 38.
Hieracium, 123.
Hippocrepis, 72.
Hippuris, 94.
Holcus, 228.
Honckeneya, 49.
Hordeum, 236.
Hottonia, 172.
Humulus, 186.
Hyacinthus, 202.
Hydrocharidaceæ, 204.
Hydrocotyle, 105.
Hymenophyllum, 243.
Hyoscyamus, 150.
Hypericaceæ, 56.
Hypericum, 56.
Hypochœris, 121.

IBERIS, 31.
Ilex, 146.
Ilicaceæ, 146.
Impatiens, 64.
Inula, 139.
Iridaceæ, 199.
Iris, 199.
Isnardia, 93.
Isoetes, 246.

JASIONE, 142.

Jasminaceæ, 147.
Juncaceæ, 211.
Juncus, 211.
Juniperus, 192.

KNAUTIA, 120.
Kobresia, 218.
Koeleria, 229.
Koniga, 34.

LABIATÆ, 159.
Lactuca, 122.
Lamium, 162.
Lapsana, 128.
Lastrea, 239.
Lathræa, 157.
Lathyrus, 74.
Leguminiferæ, 66.
Lemna, 209.
Lemnaceæ, 209.
Leontodon, 121.
Leonurus, 162.
Lepidium, 31.
Lepturus, 236.
Leucojum, 200.
Ligustrum, 147.
Liliaceæ, 200.
Lilium, 200.
Linaceæ, 53.
Linaria, 156.
Linum, 51.
Listera, 194.
Lithospermum, 167.
Littorella, 175.
Lobelia, 142.
Lolium, 235.
Lonicera, 114.
Loranthaceæ, 113.
Lotus, 71.
Luzula, 213.
Lychnis, 46.
Lycopodiaceæ, 244.
Lycopodium, 244.
Lycopsis, 169.
Lycopus, 158.
Lysimachia, 172.
Lythraceæ, 95.
Lythrum, 95.

MADIA, 133.
Malaxis, 198.
Malva, 54.
Malvaceæ, 54.
Marrubium, 165.
Marsileaceæ, 246.
Matricaria, 140.
Meconopsis, 26.
Medicago, 68.
Melampyrum, 155.
Melanthiaceæ, 204.
Melica, 229.
Melilotus, 68.
Mentha, 158.
Menyanthes, 149.
Mercurialis, 185.
Mertensia, 168.
Mespilus, 89.
Meum, 110.
Milium, 225.
Mimulus, 157.
Mœnchia, 47.
Molinia, 229.
Monotropa, 146.
Montia, 96.
Myosotis, 166.
Myrica, 192.
Myriophyllum, 94.
Myrrhis, 112.

NARCISSUS, 199.
Nardus, 236.
Narthecium, 202.
Nasturtium, 37.
Neottia, 193.
Nepeta, 165.
Neslia, 39.
Nuphar, 24.
Nymphæa, 24.
Nymphæaceæ, 24.

OBIONE, 176.
Œnanthe, 108.
Œnothera, 93.
Omphalodes, 169.
Onagraceæ, 91.
Onobrychis, 72.
Ononis, 67.

INDEX OF SCIENTIFIC NAMES.

Onopordum, 131.
Ophioglossum, 244.
Ophrys, 197.
Orchidaceæ, 193.
Orchis, 195.
Origanum, 160.
Ornithogalum, 202.
Ornithopus, 71.
Orobanchaceæ, 157.
Orobanche, 158.
Orobus, 74.
Osmunda, 244.
Oxalidaceæ, 64.
Oxalis, 64.
Oxyria, 182.

PAPAVER, 25.
Papaveraceæ, 25.
Parietaria, 185.
Paris, 204.
Parnassia, 104.
Pedicularis, 155.
Peplis, 96.
Petasites, 135.
Peucedanum, 110.
Phalaris, 224.
Phleum, 225.
Picris, 121.
Pilularia, 246.
Pimpinella, 108.
Pinguicula, 170.
Pinguiculaceæ, 170.
Pinus, 192.
Plantaginaceæ, 175.
Plantago, 175.
Plumbaginaceæ, 174.
Poa, 230.
Polemoniaceæ, 149.
Polemonium, 149.
Polygala, 44.
Polygalaceæ, 44.
Polygonaceæ, 179.
Polygonatum, 203.
Polygonum, 179.
Polypodium, 237.
Polypogon, 225.
Polystichum, 239.
Populus, 188.
Portulaceæ, 96.

Potamaceæ, 206.
Potamogeton, 206.
Potentilla, 77.
Poterium, 88.
Primula, 171.
Primulaceæ, 171.
Prunella, 165.
Prunus, 75.
Pteris, 243.
Pulicaria, 139.
Pulmonaria, 169.
Pyrethrum, 140.
Pyrola, 145.
Pyrus, 90.

QUERCUS, 186.

RADIOLA, 54.
Ranunculaceæ, 15.
Ranunculus, 17.
Raphanus, 39.
Rapistrum, 39.
Reseda, 40.
Resedaceæ, 40.
Rhamnaceæ, 65.
Rhamnus, 65.
Rhinanthus, 155.
Rhynchospora, 215.
Ribes, 97.
Rosa, 85.
Rosaceæ, 75.
Rubiaceæ, 115.
Rubus, 79.
Rumex, 180.
Ruppia, 209.
Ruscus, 202.

SAGINA, 48.
Sagittaria, 205.
Salicornia, 178.
Salix, 188.
Salsola, 178.
Sambucus, 113.
Samolus, 173.
Sanguisorba, 88.
Sanicula, 105.
Saponaria, 45.
Sarothamnus, 65.
Saussurea, 129, 248.
Saxifraga, 101.

Saxifragaceæ, 101.
Scabiosa, 119.
Scandix, 112.
Schœnus, 214.
Scirpus, 215.
Scleranthaceæ, 97.
Scleranthus, 97.
Sclerochloa, 230.
Scolopendrium, 243.
Scrophularia, 155.
Scrophulariaceæ, 151.
Scutellaria, 165.
Sedum, 99.
Sempervivum, 101.
Senecio, 137.
Serratula, 129.
Sesteria, 227.
Sherardia, 117.
Sibthorpia, 157.
Silaus, 109.
Silene, 45.
Sinapis, 38.
Sison, 107.
Sisymbrium, 37.
Sium, 108.
Smyrnium, 106.
Solanaceæ, 150.
Solanum, 151.
Solidago, 136.
Sonchus, 122.
Sparganium, 210.
Spergula, 49.
Spergularia, 49.
Spiræa, 76.
Spiranthes, 193.
Stachys, 164.
Staphylea, 59.
Statice, 174.
Stellaria, 51.
Stipa, 227.
Stratiotes, 204.
Suæda, 178.
Subularia, 33.
Symphytum, 168.

TAMUS, 204.
Tanacetum, 140.
Taraxacum, 128.
Taxus, 193.

Teesdalia, 30.
Teucrium, 161.
Thalictrum, 15.
Thlaspi, 29.
Thymelæaceæ, 182.
Thymus, 160.
Tilia, 55.
Tiliaceæ, 55.
Torilis, 111.
Tragopogon, 120.
Trientalis, 171.
Trifolium, 69.
Triglochin, 206.
Trigonella, 68.
Trinia, 106.
Triodia, 228.
Triticum, 235.
Trollius, 21.
Turritis, 36.
Tussilago, 136.
Typha, 210.
Typhaceæ, 210.

ULEX, 66.
Ulmus, 186.
Umbelliferæ, 105.
Urtica, 185.
Urticaceæ, 185.
Utricularia, 170.

VACCINIUM, 144.
Valeriana, 118.
Valerianaceæ, 118.
Valerianella, 118.
Verbascum, 151.
Verbena, 158.
Verbenaceæ, 158.
Veronica, 152.
Viburnum, 114.
Vicia, 72.
Villarsia, 149.
Vinca, 147.
Viola, 41.
Violaceæ, 41.
Viscum, 113.

WOODSIA, 237.

ZANNICHELLIA, 209.
Zostera, 209.

INDEX OF ENGLISH NAMES.

ABELE, 188.
Adder's Tongue, 244.
Agrimony, 87.
—— Hemp, 133.
Alder, 187.
—— Black, 65.
Alexanders, 106.
Alkanet, 169.
All-seed, 54.
American Waterweed, 204.
Anemone, 17.
Apple Pie, 91.
Archangel, 162.
Arrow Grass, Marsh, 206.
—— Head, 205.
Asarabacca, 183.
Ash, 147.
—— Mountain, 91.
Aspen, 188.
Asphodel, Bog, 202.
Autumnal Crocus, 204.
Avens, 76, 77.
Awl-wort, 33.

BALD-MONEY, 110.
Balsam, 64.
Bane Berries, 23.
Barberry, 23.
Barley, Wild, 236.
Barren-wort, 24.
Base Rocket, 40.
Basil, 161.
Beam Tree, White, 90.
Bean, Bog, 149.
Bearberry, 144.
Bedstraw, 115, 116.

Beech, 187.
—— Fern, 237.
Beet, Wild, 177.
Bellflower, Giant, 142.
Betony, 164.
Bilberry, 144.
Bindweed, 149, 150.
Birch, 187.
—— Black, 180.
Bird Cherry, 75.
Bird's-foot, 71.
—— Trefoil, 71.
Bird's Nest, 146.
—— Orchis, 193.
Bistort, 178, 179.
Bitter-Cress, 34, 35.
Bitter-sweet, 151.
Black Alder, 65.
—— Bindweed, 180.
—— Bryony, 204.
—— Horehound, 162.
Bladder Fern, 238.
Bladder-nut Tree, 59.
Bladderwort, 170.
Bleaberry, 144.
Blinks, Water, 96.
Blue Bottle, 132.
Bog Asphodel, 202.
—— Bean, 149.
—— Rush, 214.
Booins, 137.
Borage, 168.
Bracken, 241.
Brake, Common, 243.
Bramble, 80-84.
Briar, Sweet, 86.

Brooklime, 153.
Brook-weed, 173.
Broom, 66.
Bryony, 96.
—— Black, 204.
Buckthorn, 65.
Buckwheat, 180.
Bugle, 162.
Bugloss, 169.
—— Viper's, 169.
Bullace, 75.
Bullrush, 211.
Bull Toppins, 227.
Bulrush, 215.
Burdock, 129.
Bur Reed, 210.
Burnet, Great, 88.
—— Lesser, 88.
—— Saxifrage, 108.
Butcher's Broom, 202.
Butter-bur, 135.
Buttercup, 20.
Butter Dockin, 181.
Butterwort, 170.
Button Twitch, 228.

CABBAGE, 38.
Calamint, 161.
Camomile, 141.
—— Wild, 140.
Campion, 45-47.
Candle-wicks, 211.
Candytuft, 31.
Capon's Tail, 232.
Caraway, 107.
Carrot, 111.
Catchfly, 45, 46.
Cat Mint, 165.
Cat's Ear, 121.
—— Foot, 134.
Celandine, Greater, 26.
—— Lesser, 18.
Celery, 106.
Centaury, 148.
Charlock, 38.
—— Jointed, 39.
Cherry, 75.

Chervil, 112.
Chestnut, 187.
Chickweed, 51.
——. Clustered, 52.
—— Field, 52.
—— Mouse-ear, 52.
—— Plantain-leaved, 50.
Chives, 201.
Chicory, 128.
Cicely, 112.
Cinquefoil, 77, 78.
Closs, 212.
Cloudberry, 79.
Clover, 69, 70.
Club Moss, 245.
Cockle, 47.
Coltsfoot, 136.
Columbine, 22.
Comfrey, 168.
Cotton Grass, 217, 218.
Cowberry, 144.
Cow Parsley, 112.
—— Parsnip, 111.
—— Wheat, 155.
Cowslip, 171.
Crab-tree, 90.
Cranberry, 145.
Cranes, 145.
Crane's Bill, 60-63.
Cress, Bitter, 34, 35.
—— Penny, 29.
—— Rock, 35, 36.
—— Thorow, 29.
—— Wall, 35.
—— Wart, 29.
—— Water, 37.
—— Winter, 36.
Crocus, 199.
——— Autumnal, 204.
Crones, 145.
Crosswort, 115.
Crowberry, 183.
Crow-Foot, 17-21.
—— Water, 17.
Cuckoo Flower, 34.
Cuckoo's Bread and Cheese, 64.
Cudweed, 135.
Currant, 97, 98.

INDEX OF ENGLISH NAMES.

DAFFODIL, 199.
Daisy, 139.
—— Michaelmas, 136.
—— Ox-eye, 140.
Dame's Violet, 38.
Dandelion, 128.
Danewort, 114.
Darnel, 236.
Dead Nettle, 162, 163.
Dead Tongue, 109.
Devil's-bit Scabious, 119.
Dew Berry, 85.
Dock, 181.
Dockin, 181.
Dog Rose, 87.
—— Violet, 42.
Dogwood, 105.
Dropwort, 76.
Duck Weed, 209.
Dwarf Elder, 107.
Dyer's Weed, 40, 66.

EARTH NUT, 108.
Elder, 113.
—— Dwarf, 107, 114.
Elecampane, 139.
Elm, 186.
Enchanter's Nightshade, 93.
Evening Primrose, 93.
Everlasting, Wild, 134.
Eyebright, 154.

FENNEL, 109.
Fern, Beech, 237.
—— Bladder, 238.
—— Filmy, 243.
—— Flowering, 244.
—— Hard, 243.
—— Holly, 239.
—— Lady, 241.
—— Male, 240.
—— Marsh, 239.
—— Mountain, 239.
—— Oak, 238.
—— Parsley, 238.
—— Scale, 237.
—— Shield, 239.
Feverfew, 140.

Figwort, 155, 156.
Filmy Fern, 243.
Fir, Scotch, 192.
Flax, 53, 54.
Fleabane, 136.
Flowering Fern, 244.
—— Rush, 206.
Fog, Yorkshire, 228.
Fool's Parsley, 109.
Forget-me-not, 166.
Foxglove, 156.
—— White, 142.
Fritillary, 200.
Fumitory, 27, 28.
Furze, 66.

GALE, SWEET, 192.
Garlick, 201.
Gentian, 147, 148.
Gipsy Wort, 198.
Glasswort Samphire, 178.
Globe Flower, 21.
Goat's Beard, 120.
Gold of Pleasure, 34.
Golden Rod, 136.
—— Saxifrage, 104.
Goldylocks, 19.
Good King Henry, 176.
Goose Corn, 234.
—— Foot, 176.
—— Grass, 117.
Gooseberry, 98.
Gorse, 66.
Goutweed, 107.
Grass, Bent, 226.
—— Canary, 224.
—— Cock's Foot, 232.
—— Dodderin, 232.
—— Dog's Tail, 232.
—— Feather, 227.
—— Fescue, 233.
—— Flote, 229.
—— Fox-tail, 225.
—— Mat, 236.
—— Meadow, 231.
—— Quaking, 232.
—— Reed, 224.
—— Rye, 235, 236.

R

Grass, Timothy, 225.
—— Vernal, 224.
Grass of Parnassus, 104.
Gromwell, 167.
Ground Ivy, 165.
Groundsel, 137.
Guelder Rose, 114.

HARD FERN, 243.
Harebell, 141.
Hart's Tongue, 243.
Hawkbit, 121.
Hawk's Beard, 123.
Hawkweed, 123, 125, 128.
Hawthorn, 89.
Hazel, 187.
Heart's-ease, 43.
Heath, 143.
Heckberry, 75.
Hedge Mustard, 37.
—— Parsley, 111.
Hellebore, 22.
Helleborine, 194.
Hemlock, 106.
—— Water, 106.
Hemp Agrimony, 133.
—— Nettle, 163.
Henbane, 150.
Henpens, 155.
Herb Christopher, 23.
—— Paris, 204.
—— Robert, 63.
Hog-a-back, 119.
Hog-weed, 111.
Holly, 146.
—— Fern, 239.
—— Sea, 105.
Honeysuckle, 114.
Hop, Wild, 186.
Horehound, 165.
—— Black, 162.
Hornbeam, 187.
Horsebane, 109.
Horsenops, 132.
Horse-Radish, 33.
Horse-shoe Vetch, 72.
Hound's Tongue, 169.

House Leek, 101.
Hyacinth, Wood, 202.

IRIS, 199.
Ivy, 105.
—— Ground, 165.

JACK BY THE HEDGE, 37.
Jacob's Ladder, 149.
Juniper, 192.

KNAPWEED, 132.
Knawel, 97.
Knot Grass, 180.
Knoutberry, 79.

LADIES' SMOCK, 34, 35.
Lady Fern, 241.
Lady's Finger, 67.
—— Mantle, 88, 89.
—— Slipper, 198.
—— Tresses, 193.
Lamb's Lettuce, 118.
—— Tongue, 175.
Laurel, Spurge, 182.
Lavender, Sea, 174.
Leopard's Bane, 138.
Lettuce, 122.
Lily of the Valley, 202.
—— Turk's Cap, 200.
—— Water, 24.
Lime, 55.
Linden Tree, 55.
Ling, 143.
Liquorice, Wild, 71.
Loose-strife, Purple, 96.
—— Yellow, 172.
Lousewort, 155.
Love in a Chain, 100.
Lucerne, 68.
Lungwort, 168.

MADDER, Field, 117.
Male Fern, 240.
Mallow, 54, 55.
—— Marsh, 55.
Maple, 58.
Mare's Tail, 94, 246, 247.

Marigold, Corn, 139.
—— Marsh, 21.
Marjoram, Wild, 160.
Marram, 226.
Marsh Fern, 239.
Masterwort, 111.
May Flower, 34.
Meadow Rue, 15.
—— Sweet, 76.
Meals, 176.
Medick, 68.
Medlar, 89.
Meg wi' many feet, 20.
Melilot, 68.
Mercury, 185.
Mezereon, 183.
Mignonette, Wild, 40.
Milfoil, 141.
Milkwort, 44.
Millfoil, 95.
Mint, 159, 160.
—— Cat, 165.
Mistletoe, 113.
Moneywort, 172.
Monkey Plant, 157.
Monk's Hood, 23.
—— Rhubarb, 181.
Moonwort, 244.
Moschatel, 104.
Moss Campion, 46.
—— Club, 245.
—— Stag's Horn, 244.
Motherwort, 162.
Mountain Ash, 91.
—— Fern, 239.
—— Sorrel, 182.
Mouse-ear, 52.
—— Chickweed, 52.
Mugwort, 134.
Mullein, 151.
Mustard, 38, 39.
—— Hedge, 37.
—— Tower, 36.
—— Treacle, 37.
—— Wild, 38.

NAVELWORT, 101.
Nettle, 185.

Nettle, Dead, 162, 163.
—— Hemp, 163.
Nightshade, Deadly, 151.
—— Enchanter's, 93.
Nipplewort, 128.

OAK, 186.
—— Fern, 238.
Oat, Wild, 228.
Old Woman's Purse, 64.
Orache, 177.
Orchis, 195, 196.
—— Bee, 197.
—— Bird's Nest, 193.
—— Butterfly, 197.
—— Fly, 198.
—— Fragrant, 196.
—— Frog, 197.
Osier, 190.
Ox-eye Daisy, 140.
Ox-tongue, 121.

PADDOCK PIPE, 94.
Pansy, 43.
Paris, Herb, 204.
Parsley, Cow, 112.
—— Fern, 238.
—— Fool's, 109.
—— Hedge, 111.
—— Piert, 89.
Parsnip, Cow, 111.
—— Water, 108.
Pea, 74.
Pear-tree, 90.
Pearlwort, 48.
—— Upright, 47.
Pellitory, Wall, 185.
Penny-royal, 160.
Pennywort, 105.
Pepper Saxifrage, 109.
—— Wall, 100.
—— Water, 179.
Pepperwort, 31.
Periwinkle, 147.
Pheasant's Eye, 17.
Pilewort, 18.
Pillwort, 246.
Pimpernel, 173.

Pink, 44, 45.
—— Cushion, 46.
Plantain, 175.
—— Water, 205.
Ploughman's Spikenard, 139.
Plum, Wild, 75.
Polypody, 237.
Poplar, 188.
Poppy, 25.
—— Horned, 26.
—— Welsh, 26.
Primrose, 171.
—— Evening, 93.
Privet, 147.
Pry, 223.
Purslane, Sea, 49.

QUILLWORT, 246.
Quinancy-wort, 117.

RADISH, HORSE, 33.
—— Wild, 39.
Ragged Robin, 47.
Ragwort, 137.
Ramps, 201.
Raspberry, 80.
Rattle, Red, 154.
—— Yellow, 155.
Redshanks, 179.
Reed, Great, 226.
—— Mace, 211.
Restharrow, 67.
Rhubarb, Monk's, 181.
Ribwort, 175.
Robin run by t' dyke, 117.
Rock-Cress, 35, 36.
Rock Rose, 40, 41.
Rocket, 37.
—— Base, 40.
—— Sea, 28.
—— Yellow, 36.
Roger, Stinking, 155.
Rose, 85, 87.
—— Rock, 40, 41.
Rose-root, 99.
Rowan, 91.
Rue, Meadow, 15.
—— Wall, 242.

Rush, 211, 213.
—— Bog, 214.
—— Dutch, 247.
—— Field, 214.
—— Flowering, 206.
Rusty Back, 237.

SAGE, WOOD, 161.
St. John's Wort, 56-58.
Sallow, 190.
Salsify, 120.
Saltwort, 174, 178.
Samphire, 110.
—— Glasswort, 178.
Sandwort, 50.
—— Sea, 49.
—— Thyme-leaved, 50.
Sanicle, 105.
Saw-wort, 129.
Saxifrage, 101-103.
—— Burnet, 108.
—— Golden, 104.
—— Pepper, 109.
Scabious, 119, 120.
—— Field, 120.
—— Sheep's, 142.
Scale Fern, 237.
Scurvy-grass, 31, 32.
Sea Holly, 105.
—— Kale, 29.
—— Lavender, 174.
—— Purslane, 49.
—— Rocket, 28.
—— Sandwort, 49.
Sedge, Flea, 218.
—— Great, 214.
Self-heal, 165.
Sheep's Scabious, 142.
Shepherd's Needle, 112.
—— Purse, 30.
—— Staff, 151.
Shield Fern, 239.
Shore-weed, 175.
Sieves, 211.
Silver Weed, 77.
Skull-cap, 165.
Sloe, 75.
Snakeweed, 178.

INDEX OF ENGLISH NAMES. 261

Snapdragon, 156.
Sneezewort, 141.
Snowdrop, 200.
Snowflake, 200.
Soapwort, 45.
Soldier, Water, 204.
Solomon's Seal, 203.
Sorrel, 181, 182.
—— Mountain, 182.
—— Wood, 64.
Sowthistle, 122.
Spearwort, 19.
Speedwell, 152-154.
Spikenard, Ploughman's, 139.
Spindle Tree, 64.
Spleenwort, 241, 242.
Spurge, 184, 185.
—— Laurel, 183.
Spurrey, 49.
—— Knotted, 48.
Stag's-horn Moss, 244.
Star of Bethlehem, 202.
Starwort, 95, 136.
Stinking Roger, 155.
Stitchwort, 51, 52.
Stonecrop, 99, 100.
Stork's Bill, 59, 60.
Strawberry, 78, 79.
Sundew, 43, 44.
Sweet Briar, 86.
—— Gale, 192.
Sycamore, 59.
Syphell, 101.

TANSY, 140.
Tare, 73.
Teazel, 119.
Thistle, 129-131.
Thorow Cress, 29.
Thrift, 174.
Thyme, 160.
Toadflax, 157.
Toothwort, 157.
Touch-me-not, 64.
Tower Mustard, 36.
Traveller's Joy, 15.
Treacle Mustard, 37.

Trefoil, Bird's-foot, 71.
Tutsan, 56.
Twayblade, 194.
Twitch, 235.

VALERIAN, 118.
Vervain, 158.
Vetch, 72, 73.
—— Horse-shoe, 72.
Violet, 41-43.
—— Dame's, 38.
—— Dog, 42.
—— Water, 172.
Viper's Bugloss, 169.

WAKE ROBIN, 210.
Wall Cress, 35.
—— Pellitory, 185.
—— Pepper, 100.
—— Rue, 103.
Wallflower, 38.
Wart-Cress, 29.
—— Grass, 184.
Water-Cress, 37.
Water Hemlock, 106.
—— Lily, 24.
—— Parsnip, 108.
—— Pepper, 179.
—— Plantain, 205.
—— Soldier, 204.
—— Violet, 172.
—— Weed, American, 204.
Waybread, 175.
Wayfaring Tree, 114.
Weed, Dyer's, 40.
Weld, 40.
Wheat, Cow, 155.
Whin, 66.
—— Small, 67.
White Beam Tree, 90.
—— Foxglove, 142.
Whitlow Grass, 33.
Whortleberry, 144.
Willow, 188.
—— Herb, 91, 92.
Winter-Cress, 36.
—— Green, 145, 146.
Wood Anemone, 17.

Wood Rush, 213, 214.
—— Sage, 161.
—— Sorrel, 64.
Woodruff, 117.
Wormwood, 134.
Woundwort, 164.

Yarrow, 141.
Yellow Rattle, 155.
—— Weed, 40.
Yew, 193.
Yorkshire Fog, 228.

WORKS BY THE SAME AUTHOR.

NORTH YORKSHIRE : Studies of its Botany, Geology, Climate, and Physical Geography. *Out of print*.

A NEW FLORA OF NORTHUMBERLAND AND DURHAM. By J. G. BAKER, F.R.S., and Dr. TATE. London: Williams & Norgate. 1868.

ELEMENTARY LESSONS IN BOTANICAL GEOGRAPHY. 3s. London : Lovell Reeve. 1875.

SYNOPSIS FILICUM. By SIR W. J. HOOKER, K.H., etc., and J. G. BAKER, F.R.S. Second Edition, 22s. 6d. plain ; 28s. coloured. London : W. H. Allen & Co. 1874.

WATSON'S TOPOGRAPHICAL BOTANY. Second Edition. Edited by J. G. BAKER, F.R.S., and Rev. W. W. NEWBOULD. 16s. London : Quaritch. 1883.

FLORA OF MAURITIUS AND THE SEYCHELLES. 24s. London: L. Reeve & Co. 1877.

www.ingramcontent.com/pod-product-compliance
Lightning Source LLC
Chambersburg PA
CBHW031955230426

43672CB00010B/2153